WEIRD WATERS

THE LAKE AND SEA MONSTERS OF SCANDINAVIA AND THE BALTIC STATES

LARS THOMAS

Edited by Corinna Downes
Typeset by Prudence and the Loaches,
Cover by Jacob Rask
Layout by SPiderKaT for CFZ Communications
Using Microsoft Word 2000, Microsoft , Publisher 2000, Adobe Photoshop CS.

First published in Great Britain by CFZ Press

CFZ Press
Myrtle Cottage
Woolsery
Bideford
North Devon
EX39 5QR

ISBN: 978-1-905723-70-6

CONTENTS

INTRODUCTION

Water must surely be one of the strangest substances found on this little planet of ours. It should be in the form of - and behave like - a gas, yet it is clearly a liquid. Like all other substances known to man, it ought to contract when it freezes, but it is clear that nobody has told water this, so it expands. And then, of course, it is home to some of the weirdest, strangest and most fantastic creatures you could possibly imagine, as well as a few you would find it equally hard to imagine. Some of these are well known, and you can read detailed descriptions of them in the appropriate scientific journals. But there are also a few - actually quite a few - that so far have defied all attempts of categorisation and description. Those are the lake monsters, sea serpents, and all the other aquatic creatures almost too incredible to be real.

Aquatic monsters have a long and venerable history in the waters of northern Europe, dating at least all the way back to what must surely be the grandfather of all lake-, sea- and other monsters, the mighty Midgårdsorm or Jormungandr from the old Norse mythology – a creature long enough to encircle the globe and bite its own tail. Though some have since claimed sightings of monsters several hundred metres long, some even confusing the creatures with small islands, nothing has ever come even remotely close to the gargantuan size of the Midgårdsorm. And naturally, the only one who ever dared to challenge this monster was the old Norse God Thor – the God of Thunder.

In many ways, the Midgårdsorm is the archetypical mythological monster, but that doesn't mean all monsters are figments of the imagination. The borderline between fantasy and reality is one that is fine and fluid, with aspects of one realm carried over into the other. Real creatures can so easily clothe themselves in mythological splendour, and thus become bigger, scarier, and more fabulous.

In the centuries following the time of the Midgårdsorm, many kinds of strange creatures have been seen in the waters of northern Europe. The monsters are still very much alive, in tradition as well as in reality. And why not? After all, we do love water up here in northern Europe. Not for drinking mind you, but for pottering about in; sailing on, swimming in, fishing in, or using as a general highway on our way to do a spot of raping and pillaging across the sea. Maybe we are not so much into *that* kind of thing anymore, but we are always close to water somehow, be it the sea, a lake or a stream, and it shows – in our way of life, our way of thinking, and our way of being. The Vikings were even able to exploit the various monsters, and built their warships in their image thus enabling them to scare the pants off their enemies even before the swords and axes had come out.

The only water monster with any kind of fame outside this part of the world is the Storsjö-monster or Storsjöodjuret from Sweden. You can usually hear it mentioned in the same breath as the Loch Ness Monster and suchlike luminaries. All well and good, except for the fact that we have so much more to offer monster-wise.

This then, is the result of more than 30 years of studies in the field of cryptozoology – among dusty tomes in various libraries, in record-halls and newspaper archives, and along sun drenched, as well as freezing cold, coastlines in Greenland, Iceland, the Faroe Islands, Norway, Sweden, Denmark, Finland and the three Baltic countries Estonia, Latvia and Lithuania. It is my sincere hope that this book will entertain and perhaps illuminate; maybe even give pause to ponder on the mysteries of life. Enjoy!

A little note on the style of this tome

Many books on cryptozoology are little more than catalogues of sightings, and while they have great value for research purposes, they tend to make rather dry, perhaps even dull, reading. This – hopefully – is not the case here. I have aimed for something a bit more entertaining, and hence have only included sightings with a good or relevant story to tell. The rest of the sightings have been analysed and distilled very carefully for me to be able to present their essence, so to speak, in the book.

So many people have played a part in this book, or even supplied a part of it, that it would be impossible to mention all of them. But I do want to thank all the eyewitnesses, who dared tell me their stories, and I especially want to thank Jonathan Downes, who had the guts to actually ask me to write this book for him and his CFZ outfit. And not to mention my nephew Jacob Rask, who didn't bat an eyelid when his strange uncle asked him to make the drawings that grace this book, but just said "OK".

And remember – they're still out there...waiting!

Copenhagen – 2010

Lars Thomas

GREENLAND

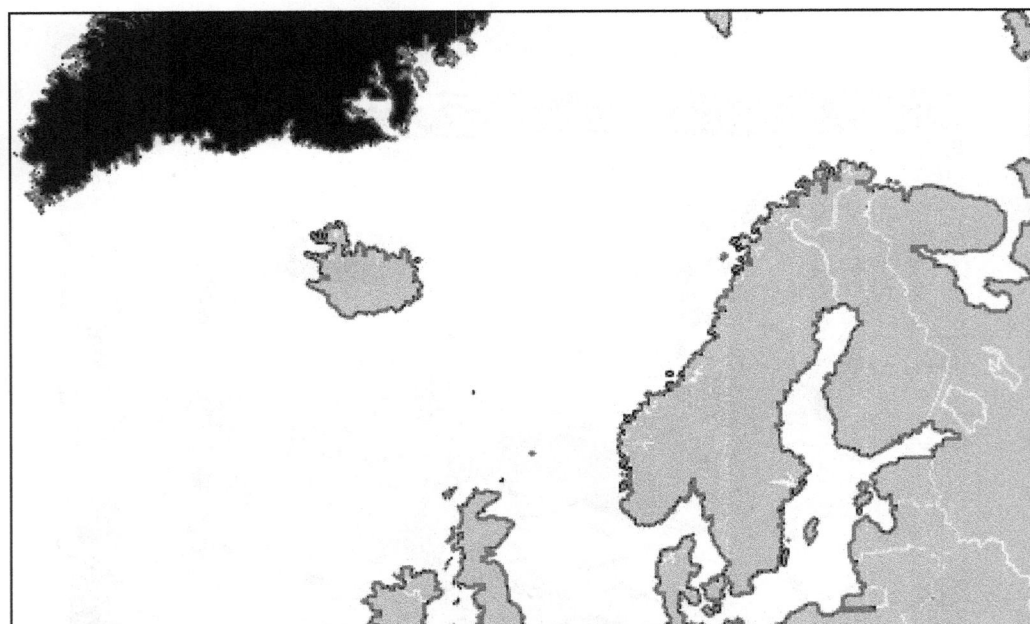

Greenland is not only the biggest island in the world (and part of Denmark) it is also one of the earliest examples of commercial flim-flam. Legend has it that the first Vikings who got there named the place Greenland in an attempt to lure settlers to follow them and start a colony, knowing full well that the place was only green for a few short months each year, whereas for the rest of the time it was one gigantic and freezing cold lump of ice. A few more generously inclined historians claim the Vikings weren't that bad – the climate was a bit more on the mild side back then, and in fact they did manage to establish a colony, although it only lasted for a fairly limited number of years.

This harsh climate means that whatever lakes Greenland has usually freeze over very quickly and thoroughly in winter, although a few very small ponds are slightly radioactive, and so usually manage to keep above freezing temperatures. Despite the relative scarcity of fresh water, the Inuits - the indigenous people - do tell a few stories of lake monsters, some of them with distinctly non-corporeal traits.

Sea monsters are not as common around Greenland as you might think, probably because the Inuits have hunted at sea for millennia, and so have a very thorough knowledge of whales and seals, and are not likely to misidentify anything. When something strange does in fact show up, you can be certain it is worth a closer look. Therefore, it is in fact from Greenland waters we have the story of one of the most famous of the lot, the Egede monster – or as the grandfather of cryptozoology Bernard Heuvelmans called it – the super-otter.

"A most terrible sea-creature"
The Danish/Norwegian priest Hans Egede (1686-1758), perhaps better known as the 'Greenland Apostle', spent fifteen years from 1721-1736 working as a missionary in Greenland. He also tried to establish a colony, and populate Greenland with convicts and soldiers. He failed though, so in 1736 he left Greenland for good and returned to Denmark.

Apart from his religious work, Egede was an eager naturalist, and he had sufficient spare time to make a substantial number of observations on the cultural and natural history of Greenland, mostly assisted by his son Poul (1708-1789), also a priest and missionary.

All these observations are noted in the books Hans Egede and his son later wrote about their life and work in Greenland. Most of the animals the Egede family describe are readily identifiable, but there is one creature, described by both father and son, that nobody has been able to identify with any kind of certainty – at least not yet.

In Poul Egede's book, one can read the following:

"On the 6th [of July 1734] a most terrible sea-creature was seen. It reached so far out of the water that its head was above the height of our yardarm. It had a long pointed snout, and blowed like a whale, had long wide flippers and the body seemed to be scaled, and the skin was very rough and uneven. Below it was shaped like a worm..."

Hans Egede describes the beast in a similar way, but his description is based on the one given by his son, as Poul was the only one of them to actually see the creature. Hans Egede was in Nuuk at the time. He notes that it was seen three days sailing north of the colony [the present Greenland capital of Nuuk], which would have put it somewhere in the Disko Bay area or in the middle of the Davis Strait between Greenland and Canada.

In Poul Egede's book we find a drawing, or actually several drawings, of the strange animal. It is quite large, probably something like 30-35 metres in length, with a long and powerful body, and a tail that ends in a point. The tail is not described in the text, but on a small picture next to the main drawing showing the animal spouting, you see the tail above the water as the animal dives.

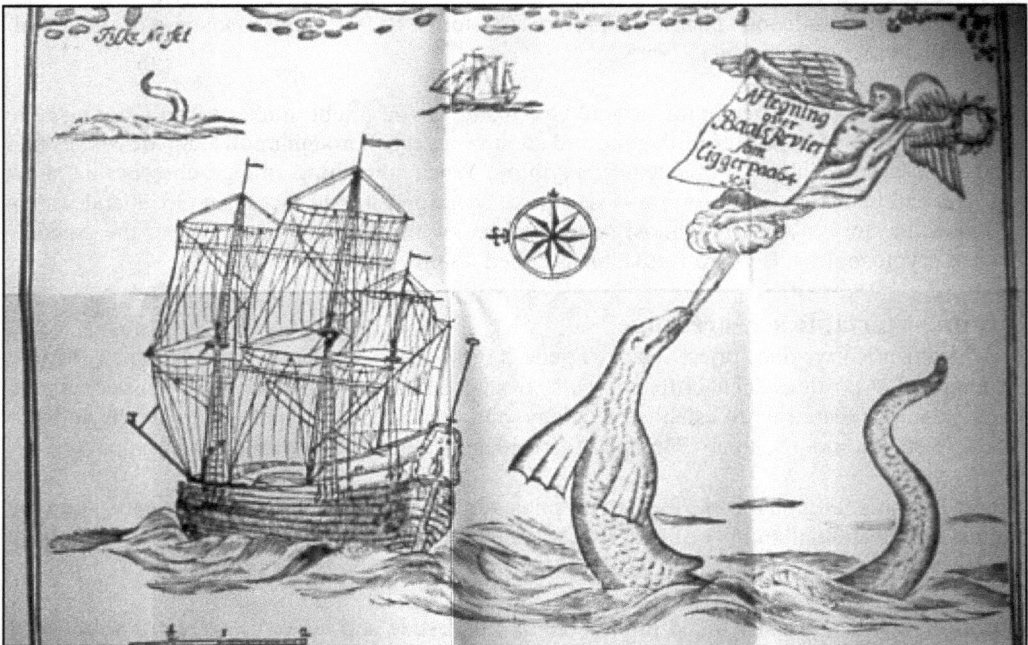

The original drawing of the Egede sea monster.
This first drawing clearly shows a small dorsal fin which is missing in several later redrawings, indicating that the animal was some kind of whale.

There is no reason to doubt the honesty of neither Poul nor Hans Egede, nor the fact that the animal Poul saw was something neither of them knew. It is not likely either, that they - for instance - misidentified an ordinary whale. A long series of drawings and descriptions of various whales and seals in their books clearly show that they were familiar with the various common sea-creatures of the area.

Present knowledge of sea-animals and the animals in and around Greenland is of course considerably greater today than it was at the time of the sighting, but we still haven't been able to nail this beast. Several researchers have tried to, but so far with no luck. Bernard Heuvelmans did publish a formal description of the creature, naming it *Hyperhydra egedei,* Egedes super-otter, as he thought it looked like a giant otter, possibly because he was working from a later drawing. In the original, a small back fin is clearly visible. In the later drawings, this fin has disappeared.

Strangely enough, although Poul and Hans interviewed the locals on many occasions about their customs and information about plants and animals, they apparently never thought to ask them about this strange creature. It might have given us valuable information. Far too often, researchers and others tend to forget that indigenous people are the best possible source of information about local animals.

The enormous size of the beast, even if you allow for some degree of exaggeration, indicates that it is some kind of whale. The flippers are typical for most whales, and the rough and uneven or scaly skin is not as strange as it may sound. Several species have large callosities in the skin, or they are covered with barnacles and suchlike creatures. This is typical for the right whales as well as the grey whale. On the other hand, none of the current species have their nostrils on the tip of the snout, as Egede's animal is shown to have. All contemporary whale species have their nostrils placed on the top of their heads. And no known whale has the same kind of long snake-like body.

Scottish zoologist Charlie Paxton and

Portrait of Hans Egede by Johan Hörner

Norwegian cryptozoologist the late Erik Knatterud have argued in a recent paper, that what the Egede pair saw, was – in the case of the "tail"- actually a male whale in a state of arousal rolling around on the surface. That may be so, but the various drawings in the books also show that Hans and Poul were quite familiar with the male whale's rather awesome "equipment". They also suggest that the animal might have been one of the last Atlantic grey whales, a species that nowadays is only found in the Pacific. The grey whales are rather long and slim animals, and might have been taken for the Egede beast, though this would mean that the whale should have lost both of its tail flukes. Another problem is that they have their nostrils placed in the "proper" place on top of the head.

So, if you stick to present species, there is no possible candidate for the Egede beast. But if you move further back in time, there is. At one time animals looking probably just as that described were found in the world's oceans.

I am talking about the now officially extinct and very primitive whale Basilosaurus or Zeuglodon. It lived, as far as we know, up until 40 million years ago. Among other places, fossils have been found along the Atlantic coast of North America. The animal was around 25 metres long, with nostrils not on top of the head but towards the end of the snout, two large front flippers and a long almost snake-like body. In most reconstructions, it is shown with tiny tail-flukes, but the tail may just as easily have ended in a point. The problem with this theory, however, is that the animal has been extinct for 40 million years. On the other hand, the coelacanth was thought to have been extinct for 80 million years when it was found of the coast of South Africa in 1938.

There is of course no real evidence for the Egede beast being a primitive whale. We have a few other sightings of a similar beast along the Norwegian coast in the middle of the 1800s, but it doesn't help us much. Unless a live individual is filmed, or a carcass floats ashore somewhere, the mystery of this "most terrible sea-creature" will never be solved.

Globsters, blobs and other rotten remains

Physical remains of sea-monsters are few and far between, though every now and then something does wash up somewhere along the very long and extremely convoluted Greenland coast. In 1984 I spent several weeks in the middle of summer on the island of Disko in Disko Bay, West Greenland. Here, in the little town of Godhavn or Qeqertassuaq, to give it its Greenlandic name, you find the Arctic Station, a research station belonging to the University of Copenhagen. I was a student at the time, mostly working with arctic birds, but I let it be known that I was interested in stories of strange and unknown creatures, and so I did manage to get a few of the locals to tell me stories about things that shouldn't be.

Every now and then very big rotten and smelly lumps - the remains of what must have been huge animals - wash up on the shores. In some cases they are readily identifiable as chunks of whale blubber or something similar, but every now and then the globsters defy identification or indeed description, as they are often just shapeless lumps with very little structure of any kind, and rarely any kinds of bones or teeth. One man told me that about twenty years before, which would put us sometime in the early 1960s, a big smelly something had washed ashore

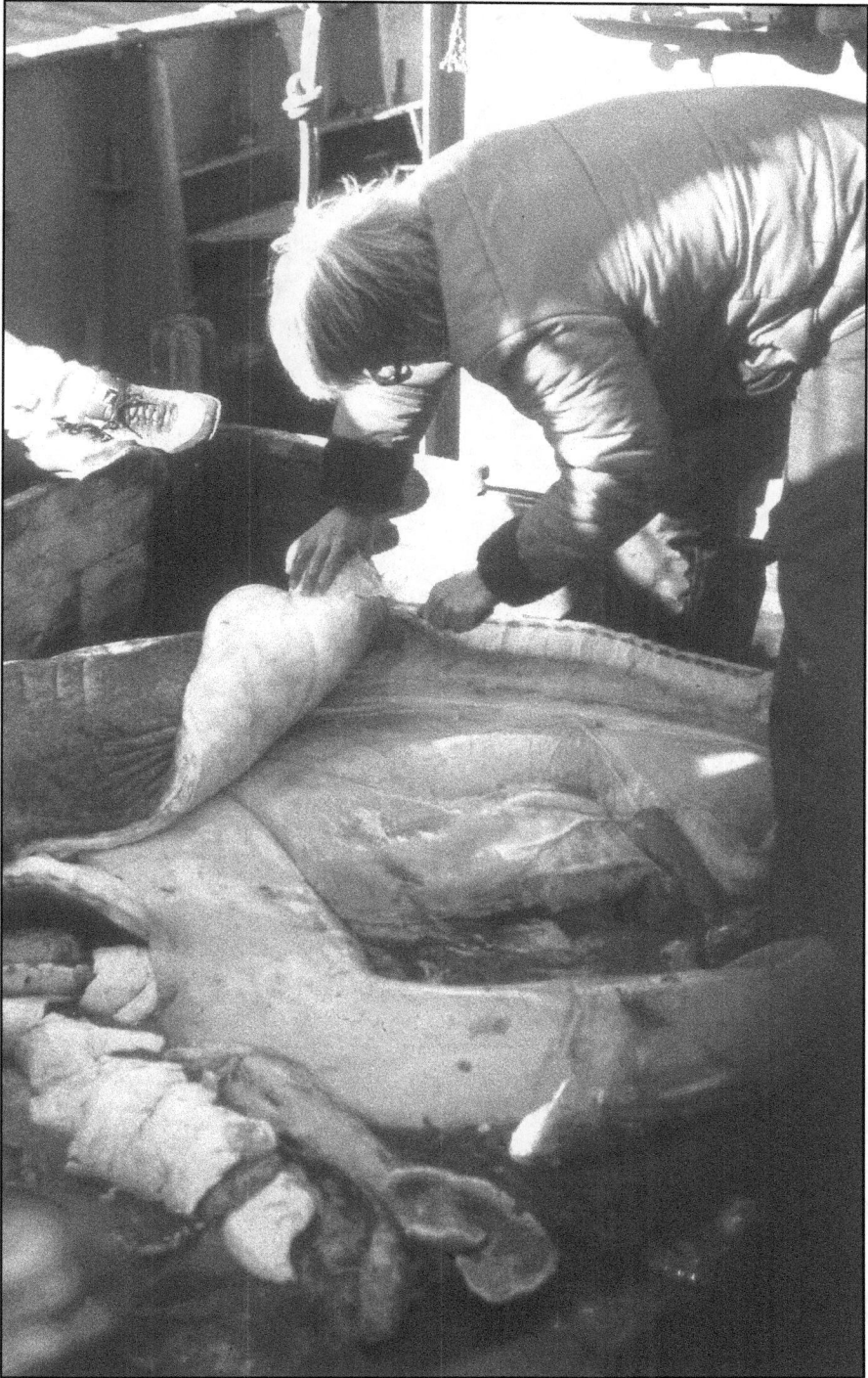

The author dissecting a dead sleeper shark - nothing mysterious was found, but food enough for an army.

The Greenland sleeper shark may be the reason for at least some of the monster sightings in Greenland waters.

on the north coast of Disko Island. Nobody, not even old experienced people, could tell what it was. It was just greyish white, flaky and greasy, with a row of what looked like gill-slits on one side. Something quite similar to that washed ashore south of the capital of Nuuk in 1992, and again somewhere around the southern tip of Greenland two years later.

An old woman described how she as a young girl, probably around 1940, had helped cut the remains of what looked like a hairy whale free of the fishing net it had entangled itself in. The creature was in a state of severe putrification, and though they saved some of the meat, they couldn't even get the sledge dogs to eat it.

Very few of these remains are subjected to any form of analysis. A piece of one lump, that washed ashore on the west coast in the middle 1990s, was studied at the Zoological Museum in Copenhagen, and was found to have come from a big shark, probably a basking shark, although the Greenland sleeper shark, which is common in the area can also grow to a substantial size.

Most of these rotten remains can probably be attributed to well-known animals; whales, big sharks, walruses or something similar, but there is no way to know for certain, unless it becomes customary to get them analysed. Unfortunately, we know very little about how large sea-animals decompose, so for the most part it is just plain guesswork.

The little Westgreenland town of Godhavn in the summer of 1984. Several strange creatures have been seen or washed ashore here, and at least one old lady in town claimed she had eaten a great auk, when she was a little girl.

Merfolk and other giants

Most sea-monsters are strange, and some are *very* strange. That really goes for the merfolk you can meet around Greenland. According to various stories and legends, there are at least three different kinds, but some of them may in fact be identical.

- The first one is the *koejak* or *auve koejak*. This is an enormous furry merman. The colour of the fur is not always described, but at least in some stories it is green.
- The second one is the *havstrambe*. This is described as having the body of an enormous fish, but the torso and head of a man with green hair and beard.
- The third one is the *margya* or *margyja*. This is a form of mermaid, perhaps the female of one of the two above. As mermaids go though, the *margyja* is not something to be excited about. It is described as being terribly ugly, with piercing eyes and a flat face.

Manatees and sea cows, and sometimes various forms of seals, are often put forward as the explanation for sightings of mermaids and similar creatures. It is quite easy to believe, when we are dealing with sightings in more temperate or tropical waters, or in areas where the local people are not used to hunting seals and other marine mammals all year round. In Greenland

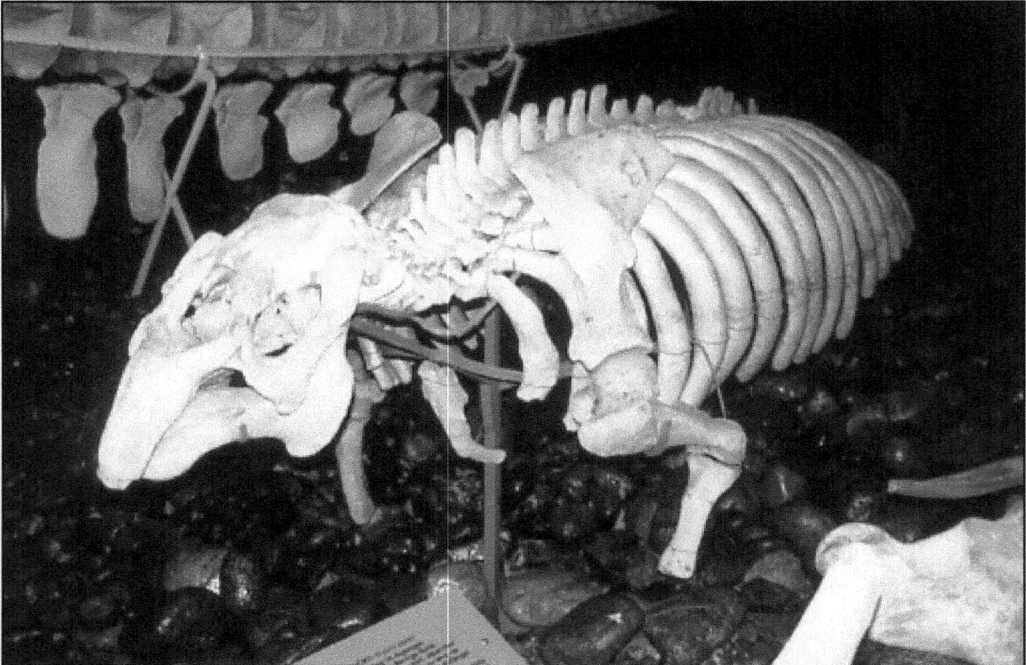

A skeleton of a Steller's sea cow, officially extinct several hundred years ago, although there have been sightings of an animal similar to that in the north Pacific and the north Atlantic.

waters, it is a bit more difficult to suggest a suitable candidate for these creatures.

One possible candidate for the first two – which are probably one and the same – is the now extinct Steller's sea cow. It was discovered and exterminated in a shockingly short time in the northern Pacific in the 1700s. It was a very large animal, but essentially just a giant version of today's far smaller tropical species. One of the things that often catches the eyes when it comes to sea cows, are the large number of brown and greenish algae growing on their skin. Having seen manatees in Belize in Central America, I can vouch for the fact that they do actually sometimes look like they have greenish fur. There is no reason to suppose that the Steller's sea cow was not similarly endowed, so this could explain the merman's green beard or fur. The only problem is that, as far as we know, Steller's sea cow has never lived in the Atlantic.

Another possible candidate is the walrus. They have the proper size and shape, but they are also very well known to the Inuits, and I find it hard to believe that they wouldn't be able to identify them properly. But in fact there are also a few stories about some kind of walrus-king, much bigger than the ordinary animals, which could be the reason for the stories. Very small or very large individuals do crop up in most animal species every now and then, and one of those, much bigger and slower than normal - thus making it possible for algae to grow on it - could be the *koejak* or the *havstrambe*. One of these could actually have been sighted off Upernavik in Western Greenland in 2003, but unfortunately, the description I was able to collect was not very detailed. The creature was seen during the summer from a motorboat, and was judged to be almost three times as long as the vessel, which would have given an animal somewhere around 10 metres in length.

The mermaid-creature must be something else. Sea cows have small roundish heads and tiny

The dreaded killer walrus. This one has fur, but it could also be found in a version with black shiny scales.

17

beady eyes, far from the flat face and piercing eyes of the *margyja*. This sounds more like some form of seal. They usually have rather large eyes and somewhat flat heads and faces, although the eyes are usually more of the large and gentle persuasion. The only thing is that the Inuit people have an intimate knowledge of all kinds of seals, so what creature could possible inspire them to tell these kinds of stories?

Maybe something along the line of a kind of killer walrus, that used to be greatly feared in the little coastal settlements. This creature has a body like the walrus, a dog-like head with large fangs, four dog-like legs, shiny black scales, and a tail like a giant fish. And this tail is actually the most dangerous part of the creature, as one blow from it can kill a man. This sounds like a mixture of several different creatures all put into one: a little bit of walrus, perhaps a dash of killer whale and a tablespoonful of shark with the legs of a wolf or something thrown in for good measure.

The final monster in this collection of odds and ends is the biggest one of the lot. The *havgufa* is so big that it is frequently mistaken for an island. When it rises to the surface, even the biggest ship will capsize. The *havgufa* is so big, it is capable of eating whales and all other sea-creatures, so a few humans are a mere titbit.

The *havgufa* must be some form of personification of the sea itself. Though stories of these colossal creatures are known from many different cultures, the likelihood of them actually being live animals is rather small. One source of inspiration could be erupting underwater volcanoes. They can create islands in a surprisingly short time, amidst steam and bubbling water, something that would probably happen faster and faster for each retelling of the story.

The weird ones
True lake monsters are few and far between in Greenland. There are many lakes – thousands of them as a matter of fact, but in most cases they are not much more than puddles usually with enough room for a few trout, but that's usually about it.

I have never heard of lake monsters from the east coast of Greenland, but from the west coast we do have some stories and sightings, and strangely enough, most of them are concentrated in the lakes around Umanak on the southwest coast. Unfortunately, most of the descriptions are rather vague, so it is difficult to know what to make of them.
In 1954, a group of women saw a huge fin in one of the lakes. It was as large as a sail, but it belonged to a large and rather ill-tempered water monster. The women didn't linger, so we don't know much more about it.

In the summer of 1963, what was probably the same creature was seen again, this time by a man hunting for ducks. This particular eyewitness was close enough to see scales on the body of the creature, and also what looked like rays in the large fin, which presumably means that the creature was some form of fish. The big-finned animal was seen again in 1977 and again in 1983, and possibly in 2006, but nothing since. One sighting describes the front end of the creature, which apparently has small beady eyes, thick lips and a fleshy tentacle or something similar in each corner of the mouth. Strange perhaps, but it could actually be the description of a big carp, or something similar.

Very large freshwater fishes are know from several areas of the world, South America and East Asia especially, but creatures like that have never been seen in Arctic areas as far as we know. So the actual identity of this creature must remain unsolved for the time being. A member of the carp family would be a suitable candidate, but we can't know for sure.

Another beast from the same general area is simply described as being large and white. That doesn't tell us much, bit it is nevertheless quite intriguing. There are not many white aquatic animals in Greenland, and especially not freshwater forms. The only reasonable candidates are the polar bear and the beluga or white whale. But both of these animals are well known, and usually living in saltwater, so we must be dealing with something else entirely. One sighting from 1969 describes the white creature as being completely shapeless, which probably means that it was in fact the remnants of a dead whale or something similar, probably washed into the lake (this separated from the sea only by a short stretch of creek). But another sighting from 1977 describes the creature as having legs or fins, and a large and rather wide head on a short powerful neck.

This white thing is definitively a fair bit out of the ordinary, but even that can't beat the *kajanok*, or as it is sometimes known, the *kajanok agdlinartok*.

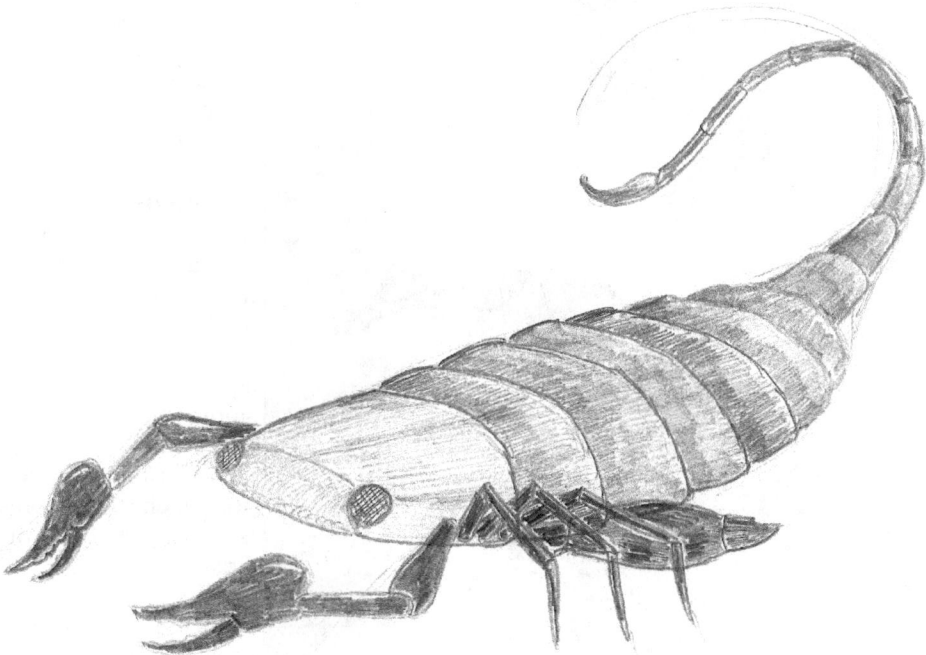

Now, the *kajanok*, apart from having a serious tongue twister for a name, is in a league of its own when it comes to sea- and lake-monsters. It is described as an enormous scorpion. Fair enough, apart from the fact that there are no scorpions in Greenland, and no scorpions ever reached the *kajanoks* rather grand proportions (it is apparently the size of a small boat), and yet the stories about this strange creature goes back for centuries. There is a strong folkloric element in the stories of the *kajanok*, but even mythical creatures are not created from scratch. They are always based on some form of real creature. So what could possibly have inspired Inuit people who had never seen a scorpion to tell stories about a creature 2 or 3 metres long, with "many legs", two "arms" armed with pincers, and a long, narrow, segmented tail?

One possibility is that we are in fact dealing with old sightings of eurypterideans or sea-scorpions, a group of animals that officially became extinct long before the dinosaurs were gone. But suppose they didn't? Suppose, perhaps, that the odd one every now and then moved inland towards the coast, or got washed in with a large wave?

What a surprise for an un-suspecting duck hunter that would be.

And the Inuit, having no written language, would just keep telling these stories again and again, never forgetting the terrifying *kajanok*.

Big birds

An interesting aside – something that doesn't actually qualify as an aquatic monster, but nevertheless concerns an at least partly aquatic animal, are some stories and sightings I have collected about a big penguin-like bird, that used to be seen on rocky shores and islets up and down the Greenland coast.

Nothing really mysterious about that, we are of course

talking about the great auk, one of the icons of extinction. The only trouble is that officially the last great auks were killed in 1844 on a little island off the south coast of Iceland, and several of the Greenland sightings are from a later date.

In 1867 great auks were allegedly seen off the coast of Disko Island. In 1888 four of the birds were seen standing on a rock in a little group of islands to the south of Disko Island called Kronprinsens Ejlande. A man I met on Disko Island in the summer of 1984 even claimed that his grandfather had caught one of these birds at some time in the 1920s. He had brought it home, and his wife had simply cooked it, and served it for dinner that night. The man I met wasn't actually born at the time, but apparently his father had told him the story.

What are we to make of these stories? I find it hard to believe that people, whose very survival depended on their ability to identify the various birds and animals of the area, would get a great auk and perhaps a razorbill or a guillemot mixed up. It is far more likely that a few scattered groups of these birds actually did survive considerably longer than normally accepted.

ICELAND

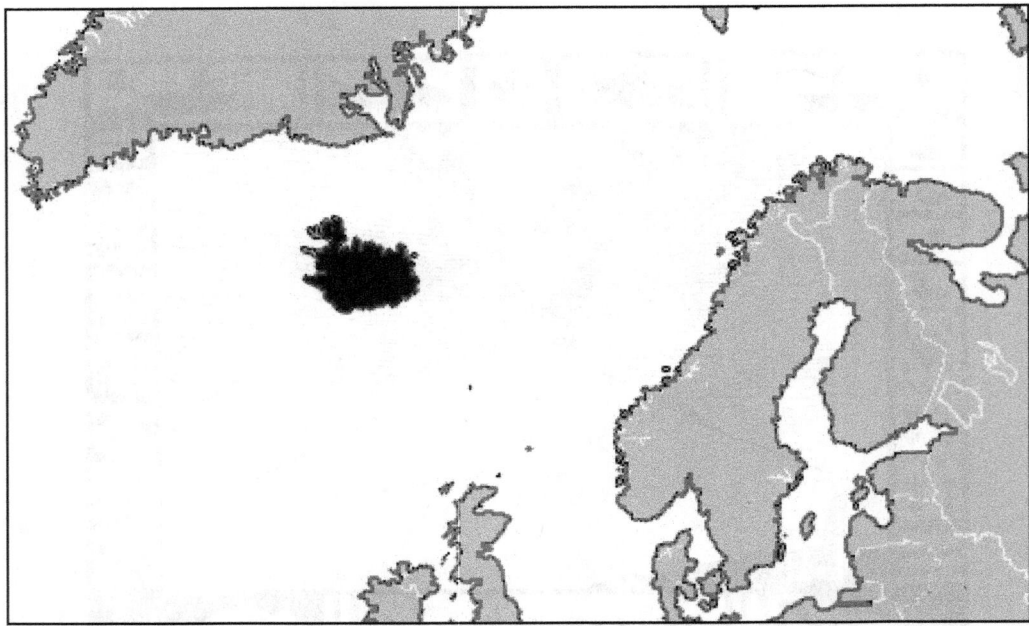

Iceland is a country of steaming geysers, volcanoes, glaciers and landscapes taken straight from other planets (no wonder a lot of American astronauts had their basic training in this country), and any self-respecting monster would feel right at home here. Some years ago, when I was stranded in Keflavik Airport outside Reykjavik, the capital of Iceland, on a very snowy December morning, I half expected to see a bunch of yetis ambling across the landing strip.

There are quite a few lakes in Iceland, some of them rather big, but they are also quite shallow. As for the sea, well – the Icelandic people are eager fishermen and whalers, so they spend a lot of time out there. And you hear quite a lot of stories about the skrimsl – an all-encompassing Icelandic word, which basically means monster – so the stories can be quite variable, presumably because several different kinds of animals are all put together under this one word. But then you also have a whole series of other strange creatures, revelling in some of the most unusual names you can possibly wish for – how about the *múshveli* or the *öfuguggi*, or perhaps the *hrosshvalur*?

I heard my first stories of Icelandic monsters at the tender age of 7, when I started in school. My teacher in Danish was actually born in Iceland, and had lived there until she was 10, but she still had a host of relatives, and access to a treasure-trove of stories and eye-witness accounts. She was my teacher for seven years, and I had contact with her for five or six years more, and during that period I collected almost 200 eye-witness accounts from her Icelandic family and friends.

The fabulous furry freak fishes

There is good fishing to be had in Iceland. Trout and salmon abound, and usually you have no trouble securing a decent catch. But sometimes you catch a bit more than you bargained for. In the summer of 1958 14-year old Olafur was out fishing in a lake called Svinavatn in northern Iceland, when he suddenly caught something big and sluggish. The fish didn't jump or fight in any way, but it was heavy and powerful and put an awful lot of pressure on Olafur's fishing rod. Despite his young age, he was an experienced angler, so after something like 20 minutes of hard work he managed to land what turned out to be a very unusual catch.

It was a big trout weighing almost ten pounds, but it was unlike any trout he had ever caught before. The fish was covered in what looked like long thin straggly hairs, like loose fur, and here and there he could also see the normal colours of a trout in bald spots where the "fur" was missing. The fish smelled strange too, so Olafur was tempted to discard it, but never having seen a creature like that, he decided to take it home hoping someone in his family knew what it was.

LODSILUNGUR

A certain amount of commotion took place in the family home on Olafur's return. It turned out he had caught a *lodsilungur*, one of a strange breed of hairy trout that sometimes could be seen in Icelandic lakes – and no, it has nothing to do with the furry trouts sold to tourists in, among other places, Canada and the US. *Lodsilungurs* are never sold. No one knew where and when the fishes would be seen – there was no apparent rhythm to it – but the weird fishes always caught people's attention. Olafur's father remembered having seen a whole flock of *lodsilungur* when he was younger, and his grandmother had actually tried to cook one when she was a girl, but the final dish smelled so strange that nobody dared to taste it.

This was probably a good idea – one has to be careful with the more strange part of the Icelandic fish fauna. After all, the hairy *lodsilungur* might be just as poisonous as the *öfuguggi*, another mysterious trout- or salmon-like being. From a cursory glance an *öfuguggi* looks like a big male salmon, but with a strange twist to it – in a very physical sense. All its fins, except the tail fin, are back to front. This gives the fish a rather bizarre look, but the worst is yet to come, for whatever you do – do not attempt to eat this fish! Don't even try to touch it, as a matter of fact. It is deadly poisonous!

According to Olafur, whom I met in Copenhagen in 1978, his grandmother also claimed that an entire family in her hometown ate an *öfuguggi* when she was a little girl, and not one of them lived to tell about it.

It is hard to know what to make of the *öfuguggi*. Big trout or big salmon is nothing unusual, and neither are poisonous fish – some species like the pufferfish are naturally poisonous, while all the rest can easily become so if you do not treat them properly when cooking them. The back-to-front fins are a bit more difficult to explain, but abnormal or malformed individuals can be found in all animal species. Perhaps the stories grew from just such a case, combined itself with various stories about fish poisoning, and ended up giving birth to the *öfuguggi*.

The *lodsilungurs* might have a much more straightforward explanation. As any aquarium enthusiast will certify, fishes do get infected with all kinds of strange parasites and diseases, including fungi, and they are quite capable of forming hair-like structures, which would explain the fur and the sluggishness of the *lodsilungur* Olafur caught, and perhaps even the strange smell. So the irregularity of the *lodsilungur* sightings might simply be explained as local epidemics of some kind of fungal disease.

The last time I talked to Olafur, in 2004 when he was 60 years old, he gave me an old drawing from his diary that he had made of the *lodsilungur* at the time, but he also claimed to have found an old photograph of the fish taken by his sister. He promised to send them to me, but I only received the drawing. The photograph never showed up. He was killed in a car crash in 2007, and although I have repeatedly asked his two daughters to look for the photograph, they haven't been able to find it.

Waterhorses and birdeaters
Vigdis, a 54-year-old dentist living in Akureyri in northern Iceland, and her son Steingrim, a 28-year-old electrician, who also happens to be a keen bird-watcher, have been my main correspondents regarding another group of lake-monsters – creatures that seem a bit more unnatural than the two strange fishes described above.

Mine and Steingrim's attempts to draw the long-necked creature he saw lying at the coast

27

One of Steingrim's favourite haunts is Myvatn (The Mosquito Lake) in northeastern Iceland. The lake is internationally known for its waterfowl. On a good day you can see almost every single European species here. It is also known for its enormous swarms of mosquitoes, not all of them of the biting variety mind you, but still more than enough to make your life miserable if you do not take proper precautions. But you can also meet something completely different.

For many years, people have reported sightings of something that looks like an upturned boat drifting around in the lake. Similar sightings have been known, of course, from quite a number of other lakes, but Myvatn is still a bit out of the ordinary as it is a very shallow lake, with limited space for any big animal. Nevertheless, something is there – and it eats ducks! In 2004, Steingrim was watching a drake Barrow's goldeneye, something of a speciality for Myvatn, when something suddenly shot up from below the surface and apparently dragged the duck under water in a flurry of foam and bubbles. According to Steingrim it happened so fast he had no time so see what actually did happen, but he never saw the drake again. He has witnessed the same thing happening twice before - once in 1996 with a group of friends, when a wigeon disappeared, and again in 1999 when disaster hit a teal.

Considering the number of birdwatchers, bristling with camera gear, that visit Myvatn every year, it is a bit strange that nobody has been able to catch the creature - or whatever it is - on film, but such happenings are rare. They happen so quickly that you would have to have your camera trained on the specific duck, and have a set of lightning reflexes to boot.

As I see it, there are several possible explanations here – one of which might be geological. Myvatn is located in a geologically extremely active area (the whole of Iceland is in a state of constant upheaval). There are volcanoes, geysers and assorted rumblings and bubbling all over the place. Maybe – just maybe – a burst of volcanic gasses are released from the bottom of the lake every now and then, rising to the surface and enveloping any birds in the neighbourhood in foam and bubbles, and probably a foul stench as well, reducing the buoyancy of the water. This could force the bird to dive and swim away, probably surfacing a fair distance away – perhaps so far away that an eyewitness wouldn't necessarily notice them.

If the birds are actually being eaten we have to assume that there is an actual animal (or several) lurking in the lake – but the question then is – what kind? The lake has a healthy population of trout and salmon, but although both of these species are voracious predators, none of them is capable of eating a fully-grown duck in a single gulp. There might be something more than fishes and ducks in the lake - there are a few sightings from Myvatn of two classical lake-monster forms: "the upturned boat" and the "longneck" known from Loch Ness and a host of other monster infested lakes around the globe.

The upturned boat, as its name implies, looks like a capsized rowing boat drifting (swimming?) on the surface of the lake. Presumably, it is the back of a large animal swimming just below the surface, and it is probably the same animal that sometimes raises its neck and head, and looks around, thus giving rise to sightings of the longneck where you see a long slim neck with a small head perched on the end.

Cryptozoologists from all walks of life have discussed the identity of these two creatures for several decades without getting anywhere near some kind of consensus, so I will not presume to put forward any kind of final solution. All I offer is a sighting that Steingrim made in 2003:

> "It was a spring day in 2003, I think in the end of May. I was in Husavik at the coast looking for gulls, when I suddenly saw a seal sleeping on the beach. There was nothing unusual about that, although I thought the animal looked very large. I thought it might be a walrus, so I started walking closer to get a better look. The animal had its back towards me, so I couldn't see the head. I thought it was asleep, but suddenly it raised its head and looked at me. It was no walrus! It had a very long neck, almost like the neck of a horse, but the head was like a normal big seal. It had big teeth. I was very scared, but so was the seal. It crawled out into the water and disappeared. I have never seen anything like it!"

What if one of these creatures made its way inland a bit, and started catching ducks?

The weird and wonderful worm

Icelandic lakes are usually either large and shallow or long, narrow, twisted and usually very deep, somewhat along the lines of Loch Ness in Scotland. This is kind of strange, or funny, or perhaps a bit of both, when you consider the fact that Iceland's most famous lake-monster bears an uncanny resemblance to Nessie in many details, right down to an early encounter with a cleric – in this case Bishop Gudmund. He bound the monster to the lake, thus ensuring that it will never leave the water, and never terrorise people on land, which is something that other Icelandic monsters do every now and then.

To encounter this fearsome beastie, revelling in the tongue-twisting name of *Lagarfljotsormurinn* - which can be translated into something along the lines of the Lagafljot Worm (Lagafljot being the name of the lake wherein the monster dwells) - you have to travel to northeast Iceland to the banks of said lake. Measuring 35 km in length, 2½ km in width, with a depth of 112 metres at the deepest, it is very similar to Loch Ness although not quite as deep.

The worm has been sighted a number of times over the centuries. The earliest known sighting is from 1345, and the most recent that I know of is from 2008. One of the most interesting sightings dates from 1998, when an entire class of schoolchildren spent almost half an hour watching a large animal swimming back and forth under the surface of the lake. Their teacher had never seen anything like it, and was not able to suggest any identification apart from the obvious – a skrimsl – a monster. The children couldn't quite agree on the shape and size of the animal they saw, but apparently it had a long neck and tail, and large flippers.

Perhaps it was the same animal seen by 17-year-old Jon in September 2008, when he went out fishing on the lake. He was just sitting there quietly, minding his own business, when a large animal suddenly poked it's head, sitting at the end of a long neck, out of the water just a couple of metres from Jon's tiny boat, stared reproachfully at him for perhaps 30 seconds, let out a strange barking sound, and disappeared under the surface again.

"I almost wet myself. I only saw the head and neck for a few seconds. It looked very much like the head of a big seal, but I had never seen a seal with a long neck. It was dark brown, with very big eyes and long whiskers."

Apparently, there are two kinds of monsters in Lake Lagarfljot, because some of the sightings I have collected tell about a long wormlike creature, although with a tendency to chubbiness around the head and neck region. There is even an old drawing showing a skrimsl looking distinctly like a snake after having worn a couple of extremely tight belts. In some cases, the creature has two large front flippers, in other cases none, and every now and then even a tail fluke much like the flukes of a whale. The descriptions vary so much that I suspect sightings of a number of different creatures have been mixed up under an all-encompassing headline. I am quite sure one could untangle all these stories, but that is almost the subject for another book in itself, so I will refrain from doing so here.

I have collected sightings from a lot of other places in Iceland, and when you put them all together you end up with a large seal-like animal, perhaps 15 metres in total length, with a long neck and tail comprising about half the animal's total length, a small seal-like head with large eyes, one or two humps, and a tendency to bark. Perhaps the very animal Steingrim saw at Husavik in 2003?

Waterhorses and mercows

Quite a number of Icelandic water-monsters seem to have amphibious characteristics. There are many sightings of apparently normal land mammals - horses and cows - walking around on land, which show an unusual degree of skill in swimming and diving when suddenly frightened or startled.

Usually these encounters are quite peaceful, but sometimes they can be rather violent. At Baularvallavatn a monster at one time went ashore and completely wrecked a farm. Information regarding the actual monster is rather scarce, so I have no idea what it looked like, but presumably it was big and angry if it was capable of that kind of destruction.

One of the most famous of these sightings dates from 1984, when two hunters, Julius Asgeirsson and Olafur Olafsson, were out hunting near Kleiforvatn, and saw what they at first assumed to be two large rocks. However, they turned out to be two horse-like animals, that at first ran back and forth along the shore, and then dived into the water, and swam away like there was no tomorrow. They were clearly not normal horses, but neither of the two men had ever seen horses swim like that, and when they examined the shoreline, where the two "horses" had been running around, they found strange footprints with three toes. Unfortunately, none of them was able to take photographs, and none of them made a sketch of the footprints, so to this day we do not know for sure how the footprints actually looked.

There are other forms of aquatic livestock in Iceland apart from these rather special horses. You can, for instance, sometimes meet up with a *sæneyti*, a kind of amphibic mercow. Stories about mercows are known from all over northern Europe. Not just Iceland, but, for example,

also in Denmark. The Danish mercows look very much like ordinary cows, although according to some stories they are a bit bigger, although some stories also claim they are smaller, and
sometimes malformed or "strange" in some way. But one thing is always certain – they give
much more and much better milk than ordinary cows. If you can find one along the coast
somewhere – possibly even a bull, and breed it with your own cows, you get the best milkers
possible. Although you have to make sure that merfolks don't suddenly show up to claim their
cow or their bull back. In Iceland, you have no doubt as to when you have seen a mercow.
Sæneyti has a strange hornlike plate on their forehead and nose. Oh, and like all mercows, they
are - of course - excellent swimmers.

This, I hear the readers cry, must surely be something taken straight out of a folktale! Most
certainly, and I would be inclined to dismiss it as such, if it weren't for the various people who
have actually claimed to have seen *sæneytis*. For instance 32-year old fisherman Jon, who on
a summer day in 2002 was out fishing almost 200 km off the northeast coast of Iceland. He
had been fishing since the age of 14, so Jon was pretty familiar with things marine. Nevertheless, he was in for a surprise. Shortly after noon, he happened to notice a group of fulmars
swimming on the surface, when the birds suddenly took flight. Something large had surfaced
among them and scared the birds away. Jon at first thought it was a seal, but then he could
clearly see a couple of horns, and a strange uneven growth, almost like the growths you can
see on grey whales or bowhead whales, was covering the animal's head. Jon at once reached
for his rifle, but the animal must have sensed his intentions, for it dived almost immediately,
but in doing so clearly showed off its horns once more.

Now what do we make of that? Frankly, I have no idea, although a walrus swimming on its
back might just give the impression of an animal with horns. I tried to suggest this to Jon
Downes a couple of years back. But he wouldn't hear of it. He knew perfectly well how a walrus looked!

The Katanes monster
This was far from the first time strange animals had been gallivanting on dry land and making
a nuisance of themselves, which is something I can sympathise with, but for some reason people never seemed to get around to actually do something about these creatures, such as shooting them, or catching them for example. Except in the case of the Katanes monster that is.

For this story, we have to go back in time a little bit to 1874 to an area in western Iceland not
far form the capital Reykjavik. It was around that time a strange creature was seen. Some local
children were the first to spot a weird dog-like "thing". It had started following them around
when they were out walking or playing, and they were at a loss as to what animal it was. They
described it as being the size of a dog, but rather elongated, with a long tail and neck, and being white or a least of a very light colour, with a reddish head.

Now what do we make of that? This was a question nobody was capable of answering at the
time, and as winter drew nearer, the creature disappeared, although sometimes you could see it
along the beach or swimming in the waters of the lake at Katanes.

Next spring the animal was back, but now it had grown to the size of a calf. It wasn't behaving aggressively in any way, and it still followed the kids around, but people were still getting worried, a worry that turned into outright fear the next year, when this strange beast had attained the size of a fully grown bull.

One of my correspondents is a man who claims that his great-great-great-great grandfather (perhaps I have forgotten a great, but I am sure you get my drift) was one of the kids, and that the story about the Katanes monster had been part of his family lore for many, many years. Apparently, his ancestor had one day actually tried to pet the creature. It didn't seemed to mind, but just lowered its neck and allowed him to scratch it. It had no ears that the boy could see, only what looked like a couple of small holes in the side of the head. It also had some kind of strange tuft on its head, hairs perhaps, or maybe some form of feelers. But it also had very large eyes, with a strange reddish colour, which makes me wonder if this was some kind of albinistic creature, red eyes being a characteristic of these aberrant animals.

Anyhow - things soon started getting ugly. The creature started attacking sheep, so the locals contacted the authorities, and asked for help. Unfortunately, nobody would take them seriously, although people were promised that if somebody would shoot the creature, they would pay for the shooting. The people hired a hunter to do the deed, but almost from the moment he turned up, the creature disappeared, and was never seen again. The hunter stayed on for quite a while, but with no luck. He didn't have much luck either when he tried to get his pay. The locals felt that since he hadn't actually shot anything, there was no reason to pay him. He actually took the locals to court, but lost his case.

Now, what are we to make of this creature? It sounds like some form of seal to me, although the long neck and tail and the white colour is rather unusual. But who knows – in Icelandic folklore there are stories about various sheep-molesting beasties, so perhaps...

Volcanic mice or something...

In the beginning of November 1963, it gradually dawned on the inhabitants of southern Iceland, that something major was about to happen. A marine research vessel cruising in the waters of the coast noticed that the seawater was considerably warmer than normal for the time of year, and people in the little town of Vik complained about a distinctive whiff of hydrogen sulphide that made their town smell of rotten eggs.

Now the Icelandic people are used to volcanic activity - they live after all in a country where things rumble, gargle and boil all the time - but this was a bit out of the ordinary. How much was clear on the 14th of November, when the cook on the fishing vessel *Isleifor II* spotted a column of smoke rising from the sea close to the Vestmannaeyjar Archipelago. This heralded the beginning of almost four years of constant volcanic eruptions that created a completely new island, later given the name Surtsey.

All this is of course well known, and it would be perfectly reasonable to ask what this has to do with cryptozoology.

2009

80

ÍSLAND

MÚSHVELI

2009

SELAMÓÐIR

80

ÍSLAND

Stamps from Iceland featuring cryptozoological animals from the oceans of Northern Europe: The mushveli is a strange whale with big rounded ears like a giant swimming mouse. Selamodir – meaning "mother of the seals" - is a strange giant seal that apparently guards the other seals.

ÍSLAND 2009

80

SÆNEYTI

The mercow with the strange nose, the sæneti, has been seen swimming far from the Icelandic coast. Skeljaskrimsl – meaning "scaly monster" – could be based on sighting of whales overgrown with barnacles and suchlike.

OPPOSITE: Hrosshvalur, the red-headed whale may in fact be garbled sightings of oarfishes. Öfoguggi, the strange trout with the backward fins is to be avoided at all cost.

80

SKELJASKRÍMSLI

2009

ÍSLAND

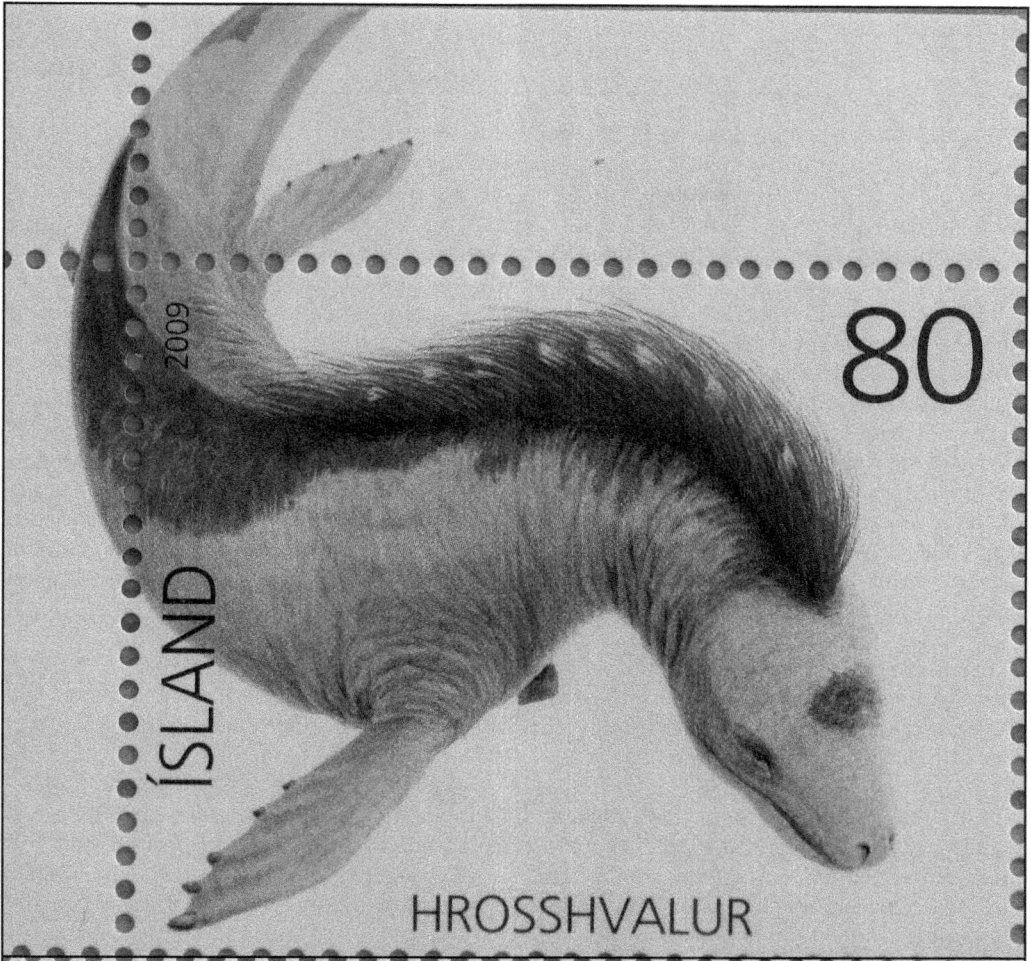

2009

80

ÍSLAND

HROSSHVALUR

80

ÖFUGUGGI

2009

ÍSLAND

Well, the thing is that all these rumblings apparently were as annoying and obnoxious, perhaps even poisonous, to the denizens of the deep as it was to the Icelandic people. Because suddenly all kinds of strange creatures started to show themselves. Petur, who is a carpenter living in Vik, and his brother Olafur, who is a fisherman, have told me a number of stories of the various beasties they saw and caught during the Surtsey eruption.

Only a few days after the eruptions started, and the island started to rise above the surface of the sea, Petur found a deep-sea anglerfish washed up on the shore, and his brother caught a whole array of strange fishes that he had never seen before. All of these were, as far as I can tell - and I was a marine biologist specialising in fishes before I became a cryptozoologist - were well known, although rarely seen, deep-water fishes that for some reason, perhaps because of the rising water temperature, had started moving towards the surface. Nothing cryptozoological about them, although Olafur said he one day caught a brick-red anglerfish that according to his description does not match any known species. It might very well have been something new for science to sink its teeth into, but he never gave it much thought, and discarded it after having shown it to a couple of friends ashore. Too bad. If only he had at least taken a picture, but at the time none of the brothers actually owned a camera.

After a few weeks things started to get really strange. Something far larger and far weirder than mere fishes started to show itself, or rather themselves, because suddenly the brothers had several sightings of something that until then they had thought only existed in Icelandic folktales.

On 7th January, 1964 Petur was standing at the harbour side in Vik, when he suddenly saw a whale jumping a bit offshore. He was used to seeing killer whales, pilot whales, the odd dolphin and sometimes even bigger whales, but this was something he had never seen before. He had no binoculars, but to this day claims he saw the animal clearly, although it was some 200 metres offshore.

The animal was long and slim, with a dirty greyish colour, a rather powerful head, and a short blunt beak. But - and this was the weird part – the whale also had a set of large roundish grey ears, almost like those of an elephant. I have suggested to Petur several times that what he thought were ears were probably the animal's pectoral fins, but he is adamant that they were ears, and that what he had seen was a *mushveli* – a mousewhale - a legendary animal described in folktales and depicted, among other places, on a set of Icelandic stamps.

Olafur, who was standing in his boat further down in the harbour heard his brother yelling about the creature, but only managed to catch a short glimpse of it as it landed in the water again, and disappeared from view. A few days later, he thought he might have seen it - or a similar creature - on one of his fishing trips offshore, but was never too sure about it.

I have absolutely no idea as to what the mousewhale actually is, unless of course it is in fact not ears, but sightings of big rounded flippers of some kind of whale – perhaps one of the many species of beaked whales that have rather large flippers. Another possibility is some form of loose skin. Most whales sometimes shed big bits of the outer layers of their skin -

whale researchers use it for DNA-analysis and similar research. It is conceivable, just, that a whale with a big piece of this skin hanging loose around its neck might look like it had big external ears. Ears are, of course, something that whales do not have, as their ears are mostly internal and they only have small ear openings.

Nevertheless, Petur and Olafur still claim that they saw a mousewhale, and Petur even managed to get their grandmother to tell a story of how she had seen one of these strange creatures when she was a little scrap of a girl, sometime around the turn of the former century - she couldn't remember the exact time - under almost identical circumstances.

The beast with the red mane

A couple of months later, sometime in the beginning of spring (the brothers have never been able to recall the exact date) the two were out fishing just for the fun of it, when they were suddenly joined by nothing less than a red-haired sea-monster. They had found a good fishing ground, and were quickly joined by various birds and predatory fishes, a dolphin or two, and something they at first thought was another mousewhale, but this creature turned out to have a long reddish mane running all along its body, from just behind the head and all the way to the tail.

Icelandic folklore has tales of these creatures. Apparently, there are two kinds: the *hrosshvalur*, which is grey, and the *raudkembingur*, which is brown, but this is the first actual sighting of one of them that I have been able to find anywhere. The animal was grey, apart from the mane, so presumably it was a *hrosshvalur*. British cryptozoologist Karl Shuker has suggested that sightings of these creatures are actually based on sightings of oarfish or king-of-herrings, an enormous pelagic species of fish with a flat, almost ribbon-shaped body - which can grow up to some 10 metres in length - and a long red back fin.

In fact, I did suggest to Petur and Olafur at the time that this was what they had actually seen, but they were quite certain they had seen a mammal. The brothers had actually once seen a small oarfish that had washed ashore when they were kids, and they were quite confident that the creature they saw catching fish among the gulls and other seabirds was a mammal. The mane was clearly made up of hairs; the animal had large expressive dark eyes, a rather elongated grey beak-like snout with lots of little pointed teeth, and a distinct blowhole on top of the head.

With all these characters in place, there is no doubt in my mind that the brothers actually saw some kind of whale – but what kind? Well, either it was a species completely new to science, or it was a deceased, malformed or overgrown individual of an already known species. But since the stories of these creatures go back quite a long way in time, it is difficult to see how some kind of defect could explain this.

Could there really be a red-haired whale out there somewhere?

Well, if only... I do have a story, unfortunately undated, and with no kind of geographical reference points that says that sometime in the 1950s a large blob of something half rotten and decayed was washed up somewhere on the south coast of Iceland, and that this "something" had a long patch of red hair on it. I have been trying for years to track that story down, but so far with no luck.

If only...

THE FAROE ISLANDS

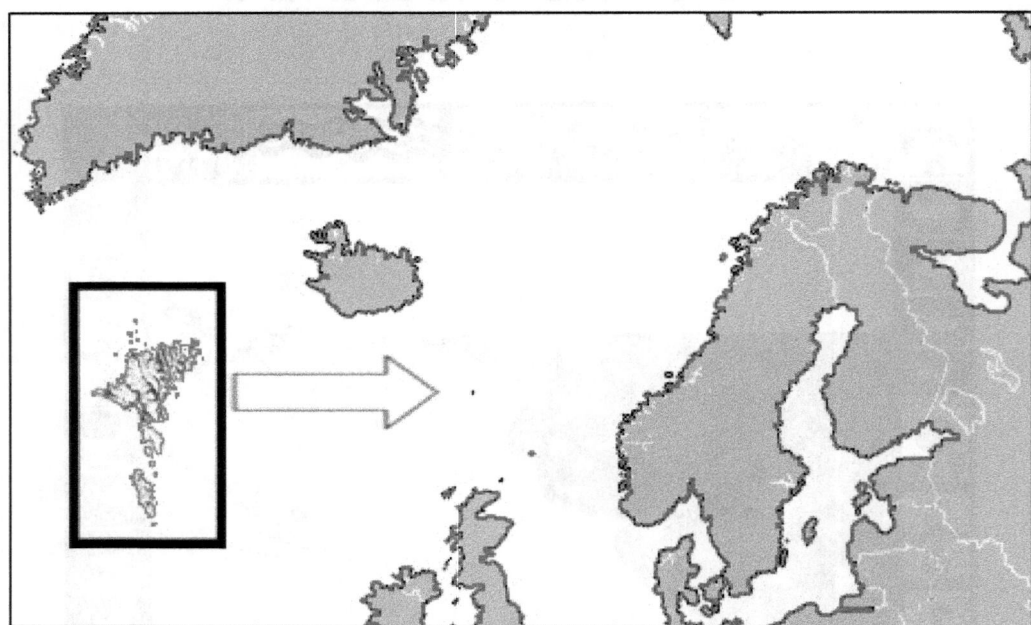

T he Faroe Islands are a part of Denmark, but located smack in the middle of the north Atlantic as the last link in a chain of islands starting with the Orkneys in the south, the Shetlands in the middle, and the Faroes in the north. The islands are small, and the sea is never far away, so there are quite a few interesting stories to be heard. Lake monsters are rare though. The islands only have a few lakes, all of them very small. Too small even for the more mysterious kind of aquatic monsters.

A horse of a different kind

Horses are not common animals on the Faroe Islands, indeed the very name of the place comes from the old Norse word for sheep, but every now and then you can still chance to meet a huge, white and quite beautiful horse. It has many different names, and relations in Iceland, Scotland and parts of Scandinavia. This is the *nivker* or waterhorse, and it is quite different from other horses. The horse - or horse, since on rare occasions you see more than one - has a rather peaceful and benign disposition. It is not especially interested in people, but doesn't seem to mind them either, so it is possible to get quite close to this creature. It is not always white though, sometimes it can be brown or greyish. The strangest thing is the creature's hooves. Apparently, they are back to front, or as one eye-witness would have it, not hooves at all, but rather webbed flipper-like feet. The reason for these rather un-horselike feet becomes apparent if one of the animals is startled or frightened. Instead of running away like any self-respecting normal horse would do, it heads straight for the nearest stretch of water, be it lake, river or the sea, and dives in. During a sighting in 1896, one of these horses even dived off a 35 metre cliff straight into the ocean.

The waterhorses haven't been seen for a long time. The most recent sighting I have been able to locate is from 1901. But perhaps it is all for the best, because apart from the rather unusual, but quite animal-like, behaviour described above, the waterhorses also have a few magical properties. Mostly, waterhorses are a peaceful lot, but they do not like to be sat upon. If you try, and are very lucky to succeed, they will only throw you, or run into the water and then throw you, but every now and then they disappear under the waves - rider and all - usually never to be seen again. Sometimes, when the waterhorses are in a particularly foul mood, they will turn themselves into either an old man with a wet beard or a fair-haired boy wearing a red cap, and actively lure people to their death beneath the waves.

To end on a strange note, legend has it that the waterhorse can also be seen in the form of a centaur-like being, with the head and torso of a beautiful young man, and the body and legs of a horse. In the summer of 1989, a Danish schoolteacher visiting the islands actually claimed to have seen a centaur late one evening whilst out for a walk. The centaur (a white one, by the way) had just stared at him, and then seemed to have faded away into the mist and the darkness. A waterhorse on the prowl perhaps?

A question of whales
The people of the Faroe Islands are famous, or perhaps infamous is the proper word to use in this day and age of environmental protection organisations and animal rights movements, for their traditional hunts for pilot whales or "grindedrab". During these hunts pods of pilot whales are driven towards the shore by people in various small boats, and then butchered when they reach the shallows. This has been going on for centuries, and thousands upon thousands of pilot whales have been killed in this manner. Pilot whales are usually uniformly glossy black, with a short beak, a rounded forehead and a small dorsal fin. But every now and then some rather aberrant individuals show up in the pods.

For five of the seven years I spent as a zoology student at the University of Copenhagen I had a friend from the Faroes. Among other things, he told me that about once a year a pilot whale with two dorsal fins would be caught. With one or two exceptions that lack dorsal fins completely, whales and dolphins normally only have one dorsal fin. Nevertheless, two-finned individuals of several different species have been sighted in various places around the globe, for instance dolphins in the Mediterranean and the Caribbean, and pilot whales (again) off the coast of New Zealand. In 1986 I saw a bottle-nosed dolphin with two dorsal fins off a remote stretch of coast in northwestern Australia. There has been some speculation as to whether these two-finned animals are new species, but considering the fact that the condition is known from several different species, it is far more likely that we are dealing with some form of rare genetic defect giving rise to a doubling of the dorsal fin. Unfortunately, none of these animals have been studied in detail, so we won't know for certain until one is caught and kept.

Another aberrant individual, in this case an animal almost completely white, was caught sometime in the 1930s. The account, which came to me from an elderly uncle of my Faroese friend, describes it as looking almost pig-like, white with a few black blotches and pinkish or red eyes, which suggests that it was in fact a case of almost complete albinism. The animal was not caught in one of the whale drives, but was found entangled in a fishing-net. It was still alive at the time, and actually scared the fishermen a bit, so they decided to release it.

The horn whale was the name given by my informant's father to yet another kind of aberrant pilot whale. This one, which he himself had seen only once, had a strange hornlike growth in the middle of its normally smoothly rounded forehead. It was only about 10 centimetres in length and looked "like a giant pointed wart", which is what it presumably was. Sightings of horned whales and dolphins of other species are known from other parts of the world, but since none of these have been examined either we cannot be certain of the cause of the "horns". But since the condition seems to transcend the borders between species, it is probably also some kind of disease or genetic defect, and not a case of unknown or undescribed species.

Out to sea
There have been quite a number of sea-monsters seen in the general area of the Faroe Islands, although none of them especially close to the coast. Most of the sightings are rather vague, only describing "a large animal" of some kind swimming just at, or below, the surface and

The foreheads of three so-called "horned whales" from the Faroe Islands. Drawings are based on eye-witness descriptions.

usually creating a large wake. A few of the sightings do stand out however. There was a sighting of a big plesiosaur-like creature about 36 km north of the islands in 1869, another of a similar creature at roughly the same spot in 1923, and finally what looked like a giant black walrus without the tusks in 1953. In 1961, what must have been some kind of baleen whale swimming on its back was seen off the east coast of the islands. This rather unusual position, also described in a couple of sightings by Bernard Heuvelmans in *In the Wake of the Sea-Serpent*, was probably due to some form of illness, although it is impossible to say for certain. Some whales, for instance humpbacks, regularly assume some rather unusual positions when in a more exuberant mood, so it might not be that unusual at all.

One of the most interesting sightings was made by Captain Brown in 1818. He had seen quite a remarkable creature. Its head and neck were upright like a mast, the body was smooth and scale less, dark brown on the upper side and white below. The animal was close to 20 metres in length and with a head some 50-60 centimetres in length. The most interesting fact was the 8 gills slits seen on the side of the neck.

Gill slits mean that this must have been some kind of fish. But which one? The creature was even bigger than the biggest known whale shark, usually regarded as the world's largest fish. The very large number of gill slits is another problem. The most primitive form of sharks only has 7 gill slits, so this creature falls outside of the normal range. Lampreys have a similar number of gill openings, although they are round and not slit-like, but even the biggest species - the sea lamprey - is only about 2 metres long. So what was this thing?

Maybe it is the same kind of creature seen in 1645 by Captain Christmas in the Danish Navy. He saw a whole school of porpoises desperately trying to escape an enormous creature resembling a gigantic swan. On several occasions, the animal lowered its long neck and small head towards the porpoises clearly trying to grab them. Apparently, it finally succeeded, grabbed a porpoise and dived, never to be seen again. This could be construed as a plesiosaur, only plesiosaurs didn't have gills – as far as we know. So what was it?

Nobody knows, but the Faroe Islands are definitely worth a visit.

The strange creature seen by Captain Brown off the Faroes - long and snake-like, but with no less than eight gill slits along the side of the neck

DENMARK

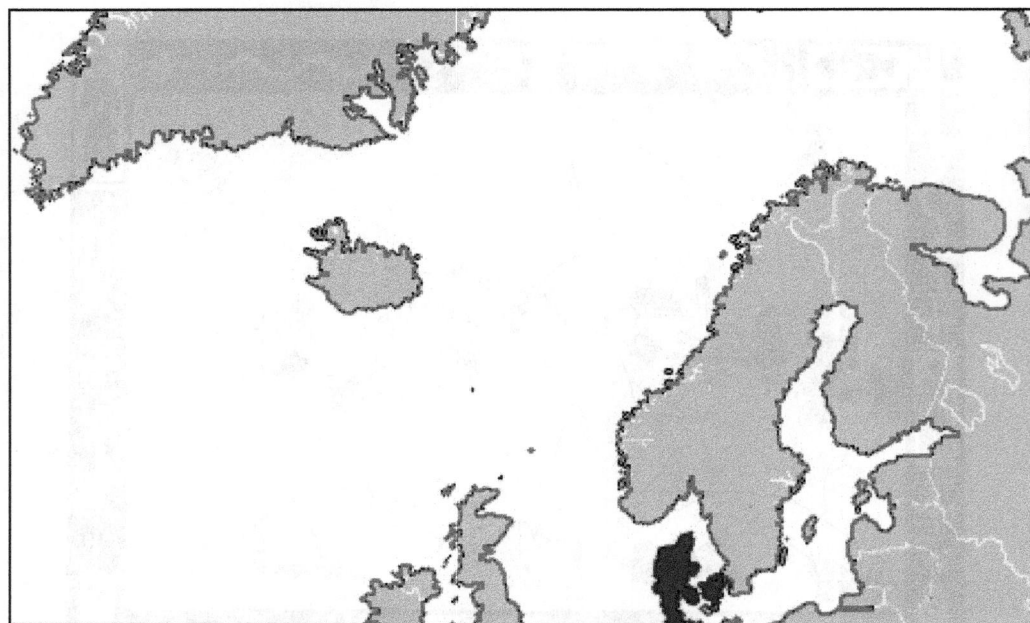

D enmark is a small and densely populated country, but with a lot of lakes and water-ways. Most of them are quite small and shallow, but there is also a lot of coastline bordering the North Sea to the west and the more tranquil waters of the Kattegat and the Baltic Sea to the east. It is perhaps not the most obvious place to go monster-hunting, but nevertheless, Denmark is home to an amazing array of strange creatures disporting themselves with gay abandon - sometimes in the middle of the capital city of Copenhagen.

Cryptozoologically speaking Denmark can also pride itself of being the home of Japetus Steenstrup, the first scientist to realise that the many stories of giant squids and sea monks were not the result of deranged and drunken sailors' imaginations running wild, but in fact observations of real living creatures – giant squids.

Small lakes – large monsters
Compared to the lakes of say Scotland, or North America, where monsters are much more common and well known, the lakes in Denmark are, with one or two exceptions, small and rather shallow. Under normal circumstances, they should not have room for a monster, let alone an entire population of an unknown species. Nevertheless, there are sightings from mill-ponds hardly big enough for a swan in which to turn around, tiny reservoirs the size of a living-room floor, and even park lakes in the middle of Copenhagen, the capital of Denmark. Who knows, Denmark *is* as mentioned above - small and densely populated - so perhaps Danish lake-monsters have to take what they can get.

Kildevældsparken, a small park just to the northeast of central Copenhagen, would seem to be just about the last place on Earth one would expect to experience anything strange or mysterious. It is a small and rather cosy little park with a lake not much bigger than the average municipal swimming-pool. It is a peaceful as you can possibly wish for on a sunny afternoon, but according to several eyewitnesses, this tiny lake is home to at least one genuine lake-monster.

The heyday of this creature was the 1970s. In those years quite a number of people walking their dog or just passing through the park early morning or early evening, especially in periods of very hot and still weather, would see something strange moving about in the little lake.

This meteorological information might seem insignificant, but this actually indicates that we might be dealing with a living creature. In hot weather the oxygen content of water drops. This means that any animal living in said water will experience breathing difficulties. In an attempt to alleviate this, the animals will move up towards the surface, where the oxygen levels are slightly higher. This naturally also makes them easier to see.

So what did people see? Most of the sightings were of something best described as a small upturned boat perhaps 2 or 3 metres long (though two eyewitnesses claimed a length of 4 or 5 metres), sailing back and forth across the lake, usually fast enough to create a small wave. These performances would last anything from a few seconds to several minutes, but all of them ended with the "boat" disappearing below the surface.

The back of a swimming animal is the obvious suggestion, or perhaps a floating tree-trunk combined with a soupçon of wild imagination. Both a distinct possibility. If it were to be a living creature, it would probably not be something large and dramatic like a dinosaur surviving from the Cretaceous. We are probably dealing with a large pike, and therefore equal parts terror and imagination must contribute towards the sighting claims of 3 - 4 metres in length. And a big pike could easily be responsible for a couple of the more dramatic sightings. Apparently a couple of unlucky dog owners letting their dogs take a swim in the lake witnessed, to their horror, the water around the animals starting to boil, and then the dogs disappearing – never to be seen again.

Most people would say that there is not much mystery about a big pike, but we are just about to enter the realm of the very strange. The monster hasn't been seen since the early 80s, but

The tiny lake in Kildevældsparken in the eastern outskirts of Copenhagen measures only a couple of hundred meters each way - nevertheless, strange critters have been seen here, and dogs have disappeared mysteriously.

The moat at Christianshavn just outside central Copenhagen is the home of a creature capable of swallowing small dogs in a single gulp, and allegedly even eat swans.

things have taken a somewhat sinister turn since then. In the summer of 1999, a young couple sneaked into the park late one night for a late and naughty swim. At the last possible moment the guy relented – he was too drunk. The girl was still game though, so she dumped her clothes and went in. And to the sound of her splashing, he fell asleep. When he awoke it was light. People were about, and one or two were actually out walking their dogs in the park. But the girlfriend? Well, her clothes were there, but not the slightest trace of her, and despite an intensive search she was never seen or heard from again.

This story has all the makings of an urban legend. I have never been able to trace the source of it, and I have since heard several variations of it from various small lakes and parks in Copenhagen. But the sinister reputation still clings to the lake, especially since a young boy actually disappeared without a trace some months later from a kindergarten just next to the park. He was found drowned in the lake several weeks later. And this time it was no legend.

Another place with a rather sinister reputation is the moat near Christianshavn in the southern outskirts of Copenhagen. Moats are strange places. Whether they surround an ancient castle or perhaps an entire town, they are always connected to strange stories and mysterious legends. Noble maidens have been pushed in, and bastard children have been drowned in them. So have useless or nosey servants and the odd gallant knight. But worst of all – moats are practically always the home of a monster of some kind. Moats are of course defensive works, so there is nothing strange in the fact that some lords of the manor have tried to make them even more formidable by stocking them with dangerous animals, or at the very least spread rumours about dangerous creatures. That is the way with the moat at Christianshavn, which is apparently the home of a dangerous monster. Maidens and knights are few and far between these days, but the monster still finds the odd victim or two.

It all started sometime in the late 1960s, when the first stories of the duck- and dog-killer surfaced. Several people had seen ducks, coots and even a swan or two disappear in a swirl of water, never to be seen again. In two or three cases, swimming dogs disappeared in this rather sinister way.

The first suspect was a large pike, which would explain most of the sightings, except the disappearing swans. A swan is too much, even for the biggest of pikes. This calls for something even bigger. And in fact there is a fish big enough to swallow a swan in a single gulp. That is the European giant catfish, up to 4 or 5 metres long, and with a weight of several hundred kilos. The only problem is that the giant catfish has been extinct in Denmark for many a year, although just a few years ago, living specimens were caught in several lakes in western Denmark, so it is not entirely impossible, although perhaps a bit implausible, that the moat was, at least at one time, the home of one of these giants.

Sortedamssøen (The Lake of The Black Pond) lies smack in the middle of Copenhagen. There is not much mystery to be had when it shimmers gently in the heat of the summer sun, and sun worshippers, people walking their dogs, and joggers each try to find a little space along its banks. But don't let the tranquillity fool you. Many a strange thing has happened here. Weird creatures have been seen in the murky waters of the lake, spaceships have landed in the middle of it in broad daylight, and some have even put down passengers.

It started in 1928, when the lake was the star attraction in the capital for a rather longish period. Two whales had suddenly appeared in the lake. In fact they were harbour porpoises, the smallest kind of whale you could imagine, but nevertheless, there they were. But where had they come from? Porpoises are sea mammals, and since the lake's connection with the sea was closed already, it is a bit of a puzzle as to how the animals had entered the lake. After a few weeks one of the porpoises mysteriously disappeared without a trace, and the other one was found dead on the banks of the lake. And that, so to speak, was that. To this day, no one knows from where they came.

About 50 years later, in 1976, things turned even more mysterious, when a strange story started to circulate in certain circles in Copenhagen. Apparently, a man on an evening stroll had seen a naked woman wading ashore from the little island in the middle of the lake. The woman claimed she had been abducted by a spaceship, and then had been brought back and thrown off the spaceship in the middle of Copenhagen.

Neither this lake, nor its two neighbouring lakes Peblingesøen and Sankt Jørgens Sø are very

When Lake Sortedam shimmers in the autumn sunshine, it is hard to comprehend, that this is one of the strangest part of Copenhagen, a place where monsters, ghosts, ufo's and all kinds of strangeness meets.

big or deep, but people claim to have seen sea-serpents here. Most of the sightings are from the 1970s and 1980s, possibly inspired by Loch Ness monster (during those years it was a frequent guest in Danish newspapers). But some sightings are plain and simple descriptions of a creature looking like a large swimming snake or giant eel, with a fin or line of long hairs along the back. The lakes have a healthy population of various fish, so some of the sightings might be put down to large pike or something like that, but on the other hand, somebody might have put something in the lake that shouldn't be there. This must have been what happened with the porpoises in 1928, so why not something else at a later date?

The strange story of the monster from Bylderup-Bov
One of the strangest stories about Danish lake monsters took place near the little village of Bylderup-Bov in southern Jutland, only a few kilometres north of the German border. During the summer of 1957, stories about a monster in a local water-filled chalk pit started to make their way to the front pages of various Danish newspapers. The stories were in a sense nothing new: stories about ducks being sucked down in a flurry of water, dogs disappearing, and sightings of the large back of something swimming just below the surface. The big difference in this case was the local blacksmith, who suggested something along the lines of "Why not try

and empty the pit?"

"So be it", the mayor probably said, and so it was. And on 23rd July 1957 a big electric pump was installed, and everybody started pumping.

The pumping plans had been described in various local and national newspapers, so a lot of curious people converged on the little village wanting to be a part of the action – although a lot of them brought packed lunches, the local baker did a roaring trade.

The locals pumped for three days, but were never able to empty the lake completely, and although they did find some rather large fishes – no pikes though – there was no sign of any monsters.

About a month later, they were ready to try again, and since they didn't find anything in the first little lake, they tried another lake, this time using an even bigger pump borrowed from a local building company. The second lake was chosen because some of the locals claimed to have seen an enormous fish in it. It was apparently a carp, but according to the eyewitnesses it was so big you couldn't reach around it. The second attempts didn't bring up any fishes, monsters or anything else for that matter. Neither did the third attempts in yet another lake, so eventually the whole thing was abandoned. The following year the blacksmith confessed that he had set up the whole shebang as a kind of practical joke – trying to make his little village famous. And in that respect he succeeded. So was it all just a hoax?

Maybe, but when I started digging into this - where you do find a lot of tiny lakes and ponds - I discovered a host of sightings of very large carp-like fishes in the area that went back at least four hundred years. There is of course nothing strange about large carp, but in each of the stories I was able to unearth, the carp - or whatever they are - are always extremely big; somewhere in the vicinity of two metres in length, and with a body as big as a barrel around. Now that is a very large carp, but on the other hand it is an average sized wels catfish, which just may be the actual animal behind all this. Wels were never a natural part of the Danish fauna, but were brought in to this country probably by monks in the late medieval period, and they just might have tried to see what could be done in southern Jutland. Unfortunately, we will never know as most of the lakes in the area have been filled in and nobody has seen the giant "carp" for many a good year now.

The sea troll and the sea monk

Sometime in the beginning of the 12th Century, when Denmark was still a Catholic country, two priests were sent to Rome to bring back a relic of a Roman martyr. Apparently, the Danish church needed some extra protection. No one knows exactly why, but according to some stories it was because the realm was plagued by an enormous beast; a sea troll with a nasty habit of capsizing any ship it saw and devouring all the sailors. Some say this monster was to be found in Storebælt, the waters between the two islands Funen and Zealand, while other stories claim it terrorised the Baltic, or even the Isefjord - a fjord you had to pass if you wanted to sail to Roskilde and visit the cathedral, the headquarters for the church in Denmark at that time.

The priests did not go to Rome with a specific order. They probably counted on God giving them a sign. Anyway, they got to the church of St. Cecilie in Rome, and here they saw a skull glowing with light like a miniature sun. This was the skull of St. Lucius, and this was a sign that everybody could understand. So he was chosen to be brought to Denmark with all speed.

The return trip went quite well, but on the last leg of the journey from Germany to Roskilde the sea troll attacked the ship. The sailors were stricken with terror, (who can blame them?) and the priests started praying to the Lord. And all to no avail, as the troll - with a suitable roar - threw itself against the ship. However, suddenly one of the monks had a brilliant idea. He opened the box with the saintly skull, and that was that. When the troll saw the remains of the holy man it backed away from the ship in terror, dived to the bottom of the sea, and was never seen again. St. Lucius you can still meet though if you visit the Catholic cathedral in central Copenhagen.

The similarity between this legend, and the story about St. Columba and the Loch Ness monster is quite striking, but as far as I have been able to ascertain there is no link between them.

Anyhow, God, or St. Columba, or Lucius for that matter, only knows what kind of creature the seatroll was. Identifying sea monsters can be a tricky business, especially in a country where ordinary monsters crop up on a regular basis. Sperm whales beach themselves on an almost yearly basis, the odd walrus shows its whiskers every now and then, basking sharks have

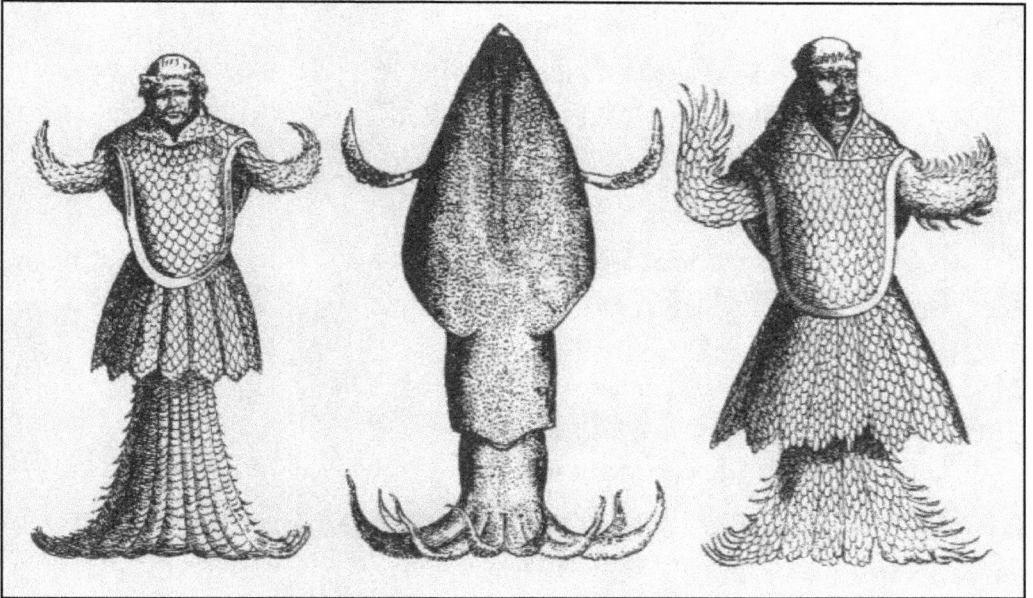

Danish zoologist Japetus Steenstrup demonstrated, that sea-monks and large squids were one and the same

passed by, and even a blue whale beached itself in the early years of the 1900s. Strange these creatures might be, they are still fairly easy to identify. The sea monks took a bit more doing, together with some help from one of the most brilliant scientific minds ever to have been born in Denmark.

The Øresund, the narrow stretch of water between Denmark and Sweden, used to be an excellent fishing area in the days before ferries and bridges. An industrious fisherman could easily bring home large catches every day. The herring was so common a pitchfork would stand upright in the water, and you could even catch a tuna or two. But every now and then the fishermen caught something very strange. In 1550 one of these gentlemen of the sea brought home the catch of a lifetime. It was a large fish, but a fish like no man had ever seen before. It was the size of a human, with a shaved head and clothing like a monk, but with scales all over, and two rather longish fins where the arms should have been.

When the fisherman got ashore and showed everybody what he had caught, nobody was in any doubt as to the nature of his catch. He had caught one of the legendary sea monks. Creatures you could read about (if you could read) in the scholarly tomes of the times. Such an important catch couldn't possibly be sold at the fish-market, so a message was sent to the king, Christian III, who at once asked to have this wonderful creature brought before him.

Johannes Japetus Steenstrup (1813–1897)

The king studied the strange creature with interest, and had it placed in the moat surrounding his castle. Here it swam back and forth very slowly, "whilst ejaculating a number of deep sighs, like a creature wrecked by the deepest sorrow and despair." The king tried to engage the sea monk in conversation, but it just sighed and swam on. The king also sent for a priest, hoping that he might be able to talk to his marine colleague, but it was all to no avail. The watery monk would not even answer when addressed in Latin.

After three days in the moat, the sea monk died never having said a word. The king was rather sorry that he never got to talk to it, but at least he made sure that it had "a good and Christian burial". Unfortunately, none of the writings of the time mentions where the monk was buried. A little spot of digging at that place might turn up something interesting.

The stories about this, and other sea monks, lived on for more than 300 years before someone finally solved the riddle of these strange beings. Were they in fact nothing but figments of peoples' imagination, or perhaps even a strange kind of hoax designed to turn the interest of the people towards the church?

In the end, it was to be a Danish scientist that conquered the sea monks. In 1853 Professor Japetus Steenstrup was told that a sea monk had become stranded on a beach in northern Denmark. Most of it had been cut up and used for dog food, but he managed to salvage a few bits and pieces, and from these he could ascertain that the sea monk was in fact a giant squid. What looked like a monk's robes was the mantle of the squid, and the scales were the pigmentations in the skin – the rest was nothing but wishful thinking. And the sighs? Probably just the water being pumped in and out of the siphon below the head of the beast as it swam back and forth in the moat. For a sea creature, swimming around in muddy freshwater around the king's castle must have been a terrible experience.

In this day and age we don't see many sea monks around here, but every now and then a fisherman drags up a hefty individual. Alas, though, there is not much dignity in being a sea monk anymore. Usually they end up in Italian stews or Spanish paellas, but the Danish fishermen still insist on calling them sea monks.

The church-wreckers

Apart from being featured in books by authors like Richard Freeman and Karl Shuker, dragons are not normally part of the curriculum when the talk turns to aquatic monsters, but Denmark is actually the home of a rather special kind of dragon with a distinct preference for water and watery places – sometimes making do with a simple well. This creature – the lindorm – has a foot in both camps, so to speak, apart from the fact that is doesn't actually have any feet. It resembles a giant eel, or perhaps an enormous snake, with a large head, large eyes and fang-like teeth, and perhaps some kind of loose-hanging frill on the neck and back of the head that sometimes stretches some way down the back. The snake ID is perhaps the most likely, since quite a lot of old stories and folktales about these creatures mentions some form of shedding of skin, as well as the fact that a lindorm seems to be as equally at home in the water as it is on land.

This fountain in front of the Town Hall in Copenhagen shows a lindorm and a bull locked in their final deadly battle. The lindorm is a bit on the short side - there is no way it would be able to reach all the way around a church. But I suppose a bit of artistic license is permissible for at fountain.

A lindorm has a rather strange biology (which brings the basilisk to mind, so perhaps the two are related in some way). It starts out as a tiny worm-like creature that may appear in a midden or a compost heap. If it is not killed outright it will escape down a well, a deep dungeon, or somewhere similarly dark and damp. Here it will grow, and grow, and grow, and grow, sometimes for centuries, and usually reaches a fairly respectable size. Should anybody, or perhaps even some careless form of livestock venture near, he or she will be promptly eaten.

The strangest thing is that according to legend quite a number of lindorme (lindorme is the plural of lindorm in Danish) have a distinct dislike of churches. When they get big enough, and if they get the chance, they will curl around the nearest church they can get to, keeping the churchgoers from entering, and seek to destroy the building. This will happen if they grow big enough to reach all the way around and grab their tail in their mouth. (Shades of good old Jormungandr (the Midgårdsorm), that reached all the way round the Earth, and grabbed its own tail.)

Quite a few gallant knights have tried to kill a lindorm, but few have succeeded. Apparently, the only sure fire way to kill one is to raise a bull on milk only, which will make it grow to an enormous size, and give it enormous strength. When such a bull is let loose, it will kill the lindorm and save the church, but be killed in the process as well.

All the folkloristic trimmings apart, there are actually quite a few sightings of lindorme from Denmark. Nørre Søby Lake, about 10 km south of Odense on the island of Funen in central Denmark, is the home of a lindorm of the worst kind, with a penchant not only for wrecking churches, but any old building it can sink its fangs into. Nevertheless, quite a number of people claim to have seen a lindorm disporting itself on the surface of the lake. In all cases, it has been snakelike, with lengths varying between 2 and 4 metres - rather modest as lindorme go - but not too mysterious. Large sterile eels, which have never gotten around to taking the long and arduous trip to the Sargasso Sea, might just be the explanation in this case.

There are also some accounts of lindorme with lengths of 30 metres, which is somewhat in the neighbourhood of a fully-grown blue whale. I do feel, however, that by adding the length of the animal's wake as well the size of these creatures may have been slightly exaggerated.

The sightings of these eels, or snakes, or worms, or whatever, are dotted fairly evenly all over Denmark, most of them rather alike though a few of them stand out. In 1943, a man in central Jutland (western Denmark) claimed that a giant snake attacked him, but that he managed to beat it off with his walking stick. And later the same year, another man - this time in south-eastern Denmark - told a local newspaper that his dog was attacked by a giant snake-like creature that tried to grab on to the dog's back leg. Well ... eels are capable of land-crossings, sometimes for quite some distances, so perhaps...

Running water and rising monsters

In most books on aquatic monsters, creatures from rivers and streams are often conspicuously absent, and you get the distinct impression that strange watery creatures are to be found only in seas, oceans and lakes. Until fairly recently, I would have said nothing different, but - although I knew a few Danish legends regarding monsters in rivers - much to my surprise when I started digging a little deeper, I turned up a number of sightings of some very strange creatures indeed.

One of them brings to mind Armus, the malevolent creature from the *Star Trek Next Generation* episode "Skin of Evil". The only difference is that this creature was seen quite a number of years before this famous TV-series was filmed.

For a meeting with this creature, we must travel to western Denmark, to Alling Å (Alling Creek) about halfway between the two towns of Randers and Rønde in eastern Jutland. If you make a stop today on the bridge, you are in the middle of a typical Danish landscape, except for the large upright stone on the meadow west of the creek. You can easily see it from the road, but if you get up close, you can see the troll-like face carved into the stone. This is the creek guardian.

The author's drawing, based upon eyewitness testimony. According to the eyewitness, this strange being rose from the river, and after a few moments sank below the water again

The stone carving is in fact from the Viking age, around 1000 A.D. At that time, it was probably painted in loud colours, but the actual purpose remains unclear. Perhaps it was supposed to keep all evil away from what has been a ford for many centuries. Perhaps it is a picture, and perhaps a warning, of the creek-man. This creature, about which you can find various legends all over Denmark, was usually found in creeks with bridges and fords, or perhaps around ferries. Generally he was no trouble, but once a year he demanded a sacrifice, and should he fail to get it the local people would pay a terrible price, perhaps in the form of a whole group of children drowning simultaneously during a crossing.

The stories about the creek-man smacks of "Bogeyman legends" - stories told to keep children away from dangerous places - but one Danish woman actually claims to have seen the creek-man.

One spring day in 1963, a 22-year old woman from Randers was standing on the bridge looking out towards the big stone, when she suddenly saw a dark shape rising from the water, like a person standing on an elevator platform. It looked like a man dressed in a black blanket. Nothing much happened, and the young woman was just about to try to get closer when the black thing just sank under the water again and disappeared. A rather large figment of the woman's imagination? Perhaps, but when she told the story in 2002 she was adamant about the details of what had really happened.

I must confess, I haven't got the faintest idea of what this creature could possible have been. A few of the stories of the creek-man from other parts of Denmark describe him as being black and rather shapeless – perhaps the same thing, or perhaps some form of manifestation of an elemental aspect of nature.

Things are a bit easier to explain if we move to Storåen (Big Creek) that runs through the town of Holstebro. A lindorm used to live here, but was killed sometime during the early Middle Ages by a knight returning home from the Crusades, but since then there have been several sightings of big snakes in the creek. In 1948, 1949, 1952, 1959, 1965, 1981, 1992 and 2001 various people claimed to have seen big snakes swimming in the creek, and in some cases they might have been exactly that. It wouldn't be the first time a pet that had turned out to be to difficult to handle had been released into the wild without any second thoughts – but continuously for more than 50 years? Oh, and people have seen crocodiles in the creek too...

NORWAY

For some strange reason, Norway is home to far more than its fair share of aquatic monsters. Pick any lake in Norway, and you are more than likely to hit on a treasure trove of stories and sightings of unexplained creatures. And you will be hard pressed to find a coastal village where the people haven't at one time or another seen, found, or caught a sea monster. Although not all of them are impressed by it – when on a visit to Trondheim I asked a local how I should go about it if I wanted to see a monster, he suggested I tried looking in a mirror!

The wealth of material, and perhaps a dash of linguistic barriers, have ensured that the Norwegian stuff is also ripe with mistakes and misunderstandings. For instance in quite a number of books on monsters, you can read about the monster of Farrisvannet. In some books you will find this placed under Denmark, which technically is partly true, as Norway was once part of Denmark, but now it is a sovereign country so please try to remember that in the future. Furthermore, Farrisvannet is not actually a lake, but a fjord. Vannet means "the water", so Farrisvannet is "The water of Farris". A lot of Norwegian lakes are called something - vannet, and

Sea monsters have in some cases been seen just outside the entrance to Trondheim harbour.

quite a lot of these are actually fjords, so the creatures found in them are more sea- than lake-monster. And, of course, there have been some rather – how to put this in a delicate and diplomatic way – ridiculous cases of persons and organisations doing their very best to confuse the issues even further, all the while claiming they were the only ones who actually knew anything about Norwegian lake-monsters.

Anyway, the Norwegians have a long and time-honoured tradition of being expert seamen, fishermen and whalers, some of which have ensured them a certain notoriety even to this day, where Norway is one of only a handful of countries that insists on maintaining a whaling industry. But – ethics aside – all this has ensured that even the average Norwegian has a higher than average knowledge of the sea and its inhabitants, so if a Norwegian claims to have seen something strange in the water you can be pretty sure that is exactly the case.

Now, Norway is one of the classic locations for one of the classic sea-monsters - the mighty and notorious Kraken, which the Norwegian bishop Erik Pontoppidan described several hundred years ago. Much has already been written about the Kraken, for example how it was extremely large (and if you do not believe me, just go and watch *Pirates of the Caribbean*!) and full of arms. In addition, how you would always know when the Kraken was near because the bottom of the sea would suddenly be much closer than it should be, and there would be scores of fish around, so I will not go into details about it. However, one or two sightings and beliefs are worth mentioning. Many people believed, for instance, that amber was in fact the excrements of the Kraken, but please do not ask me to explain why.

There have been quite a lot of sightings of Krakens over the centuries, and if you go into detail it is quite clear that we are dealing with two different kinds of "creatures", or phenomena. There are the stories of how sailors mistake a Kraken sleeping at the surface for a small island. They go 'ashore', perhaps even light a fire, and end up in dire straits when the creature awakes, and takes issue with the fact that someone has started using it's back as a barbecue, and – naturally - dives underwater with all due haste, usually dragging the men, and sometimes their boat with them.

Much has been written into these cases, with sceptics claiming that no animal could ever grow that big, whereas believers have maintained that this only shows how little we know about the inhabitants of the sea. Well yes, that much is true, but there are limits to what one can comfortably believe. It is always a good idea to keep an open mind, but not so open that your brains fall out.

These Kraken stories are undoubtedly cases where geology and oceanography, rather than zoology, should supply the explanation. As Oll Lewis from the CFZ so convincingly demonstrated at the 2009 Weird Weekend, during exceptionally low tides, reefs and perhaps sandbanks could easily be exposed so that one would suddenly see what looked like land in areas where no land had ever been seen. "This surely must be an animal," people would cry when

OPPOSITE: Even some of the carvings of the Nidaros Cathedral in Trondheim speaks of how common sea monsters are in Norway.

this newly found land suddenly disappeared under the waves again, but they did not take into account how quickly the tide can change, especially when experienced from something very low-lying.

But there are other Kraken stories, which clearly demonstrate that some stories are descriptions of giant squids that, for some reason or other - probably sickness, injury or something similar - has moved to the surface, or been washed ashore. There are numerous cases, not only from Norway, but from all over the world, of giant squids washing ashore. In most cases, the animals are dead or dying when this happens. Actual sightings of these enormous squids with some semblance of life in them are extremely rare. It was only a few years ago when Japanese scientists managed to take the first ever live footage of a giant squid in full flight so to speak, and in its natural element. Nevertheless, some people have actually seen a living and active giant squid, or a Kraken, if you like, and at least one of those lives in Norway. And her story has, to the best of my knowledge, never been published before.

I met Maria almost 20 years ago, on a sunny summer's day when I was taking pictures of a set of mosaics outside the Town Hall in Oslo. These mosaics were all depicting merhorses and other strange creatures, so my mind was very tuned into strange beings, when suddenly this tiny blond woman, standing only about five foot tall, suddenly materialised - or so it seemed - beside me, and asked what I was doing. I told her of my interest in strange creatures, and she lit up in a big smile, and said that in that case she had a story to tell if I had time to hear it. We

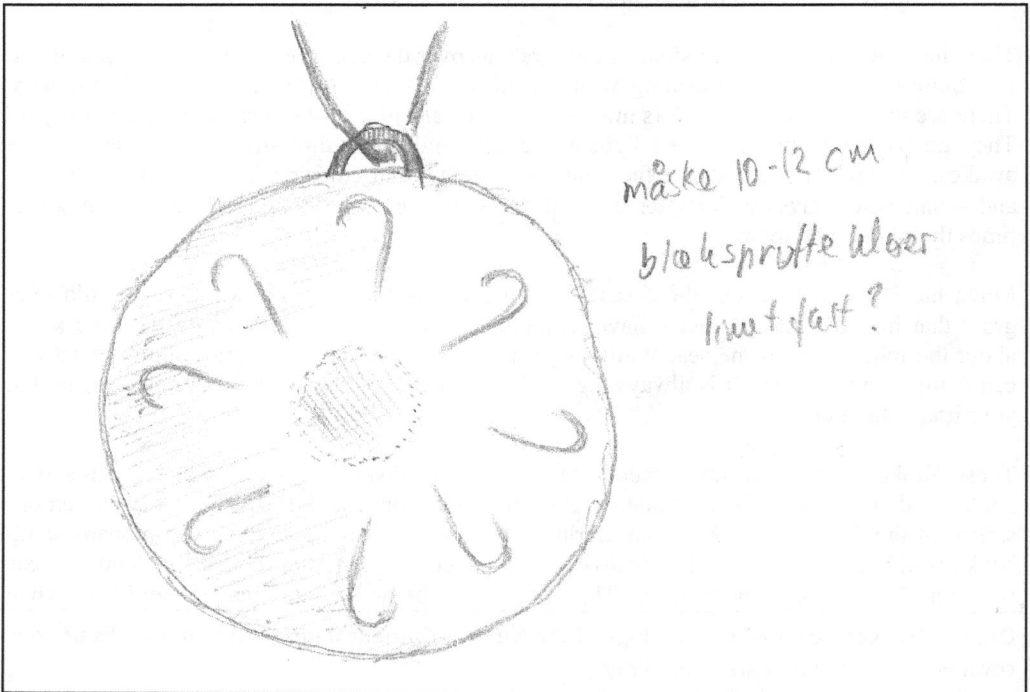

måske 10-12 cm
blæksprutte klæer
limet fast?

found a nearby café, and Maria, who at that time was a student at Oslo University, took off a medallion she was wearing around her neck, and showed it to me.

At first glance it was nothing special, just a round, rather rough brass-plate with a strange pattern of brownish lines, almost in the shape of question-marks. But then it suddenly dawned on me that the lines were not some form of artistic decoration. They were the pointed hooks that you find in the suckers of giant squids. Now I was impressed!

"I come from Haugesund on the west coast, and in the summer of 1976, when I was ten, I went out sailing with my uncle. It was an extremely hot summer, and the sea had been very quiet for days, so we spent a lot of time sailing, as it was nice and cool on the water."

"It was a day in the beginning of August, I can never remember exactly when, but we went out as usual. The sea was like a mirror, and I was just lazing about, when I looked over the side, and seemed to see something big and light moving below the surface. It was just a big whitish blob, but after a few moments, it seemed to rise towards the surface, and rolled over on its side, and suddenly I could clearly see that I was being watched by the biggest eye I had ever seen. It looked to be the size of a dinner-plate. I wasn't scared in any way, just completely mesmerized."

"I have no idea how long I watched this creature, but suddenly a long snakelike arm rising up from below very slowly, hit the side of our boat, and then started to, well, I can only describe it as feeling the side of the boat, just the way my blind cousin's hands moves, when I give him an object he hasn't touched before. Suddenly I realized my uncle was standing next to me, but he looked absolutely terrified, and told me to sit completely still."

"I still don't think we were in any danger, I think the creature was just curious, but when my uncle suddenly started the engine with a roar, he must have startled it, for the arm disappeared below the water, but two seconds later came up again like a rocket, grabbed the side of the boat again, just below the spot where I was sitting, but this time I had a glimpse of something that looked almost like cat's claws gripping into the timbers of my uncle's boat. And then my uncle suddenly brought the boat full speed ahead, and the arm released its grip with a strange velcro-like tearing sound."

"My uncle didn't stop until we were safely back in the harbour. I thought it had been extremely exiting, but he was white as a sheet. Nevertheless, next Christmas he gave me this medallion with the claws he had removed from the side of his boat a couple of days later. He said it would protect me from any further attacks. And maybe it works, because I have never seen anything like it again."

Weird waters

Fascinating creatures though squids are, at least we know they exist, and we can be certain that someone, someday, will make a proper study of them. But some of the marine creatures that frolic along the coasts of Norway still defy explanation, because these are truly weird waters. There is an amazing amount of sightings – more than enough for a book in itself - but I will just present a few of them. Some sightings were made far out at sea, off the coast, but there have also been a lot in the deep and narrow fjords along the west coast. There are apparently several different species of giant sea-creature to be seen along the Norwegian coast. The most common one is long and slender, almost snake-like, with a rather elongated head -horse- like perhaps – and a rather pointed snout. And then there is a rather more substantial one, often with a big dorsal fin or a hump. Fishes? Whales? Giant marine snakes? Well...

In August 1964 three people saw an 11 metre long sea-serpent in Kragerøfjorden in southern Noway. It was "horrible to behold", as the newspaper account would have it, and of a weird bluish-greenish colour. It was seen by a young girl in a motorboat, and later on by a fisherman and a boy in a rowing boat. It was, according to their testimony, very snake-like in form and movement, although whether it actually undulated sideways, indicating a reptilian or piscine ancestry, or vertically, which would indicate is was some kind of mammal, nobody bothered to ask the eyewitnesses, and I have not been able to track them down. Something similar, although brown, almost twice as long, and with a dorsal fin, was seen in Hessafjorden in June 1999, and again in 2001. Sightings in this area are so common that the creature(s) have their own name – "Hessie".

Hessie, as well as several other creatures of similar bulk and ilk, has also been seen eating from the carcasses of dead whales. Whether they have killed the whales themselves, or are just scavenging nobody knows, but could we perhaps be dealing with garbled sightings of sharks or killer whales? All of which are more than happy to partake in such a gory feast.

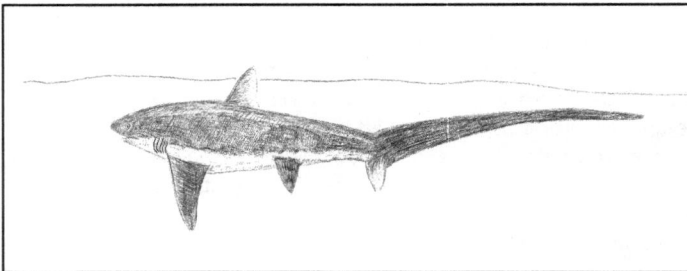

Most of these are not particularly snake-like, but I am wondering if perhaps some of the "snake" sightings are based on thresher sharks? They have extremely elongated upper tail-flukes, which would presumably look very snake-like if the animals were swimming just below the surface. I have only seen a live thresher shark once, so I can testify that the weird tail really moves in soft snake-like curves. The only problem is that threshers do not grow to 20 or 30 metres.

And what are we to make of this one?

Early one morning at the end of February 1933, the people on board the yacht *Tommy*, which was sailing in the Oslo Fjord, saw a strange creature swimming east of the yacht. The sea was calm, and the creature was passing fairly close by, so everybody had a good look for almost 10 minutes. The animal's horse-like head was reaching about a metre out of the water, and was swaying from side to side, presumably in turn with the animal's swimming movements. Three black and shiny humps could be seen behind the head, altogether measuring some seven metres, but judging from the movement in the water, the animals was at least twice as long. A giant seal perhaps?

The mysterious merfolk
In most other parts of the world mermaids, mermen and merfolk in general are a well-known but generally extremely rare phenomenon. Not so in Norway, where merfolk are, or rather were, extremely common. They have been seen, caught - and in one case boiled down for the oil in their liver - since medieval times. Some sightings have been made by people from the highest rungs of the social ladder, and a lot of them by fishermen and other members of the general populace. There are literally hundreds of these sightings, but I will only describe a few of the more interesting ones here – mainly those that would be called (were we talking about UFOs and aliens) close encounters of the third kind (or even higher!).

What is perhaps the most telling about those sightings is the very matter-of-fact style in which they are presented. For most of the people it seems that meetings with merfolk are not as strange as you might think. Although not everybody treats merfolk in as businesslike manner as Peter Angel, when he found a dead merman on a promontory off Ulstahoug in northern Norway following a violent storm in 1719. There were many other dead sea-creatures washed up, but the merman was a bit out of the ordinary. He was described as being large, some 5-6 metres, dark grey all over, with the lower half of the body being fish-like, although with a tail like a porpoise, and a human face with a very flattened nose. The arms were attached to the sides of the body with thin layers of skin, almost like the flaps you would see between the legs and the body in flying squirrels, and the hands looked like the hands of a seal. Despite his rather interesting find Angel was not impressed, at least not enough to call in science. Instead, he butchered the merman, "which yielded a good amount of blubber, and whose liver gave several litres of good quality oil". Other merfolk have been found dead or have been caught before, as well as after this particular incident. The earliest one I have been able to find any description of is from the end of the 1200s although not very detailed.

One of the more strange cases took place in 1619, when a ship returning to Copenhagen from Oslo caught a merman just off the mouth of Oslo Bay. On board this vessel were two of the king's counsellors, Ulf Rosenspore and Christian Holk. They were very interested in the merman, probably entertaining plans of showing him to the King in Copenhagen. Unfortunately, the merman didn't agree, and had no wish to meet any kings, Danish or otherwise, although apparently he could speak Danish. That was the language he used when he threatened the two noble gentlemen with storms, killer waves and general disaster and mayhem, if they didn't release him at once. This they duly did and continued their onward journey without any mishap. And the merman disappeared with all due haste.

Some 350 years later, the Norwegian merman was still in a rather foul mood, although his linguistic abilities had deteriorated somewhat. In August 1967, fisherman Kåre Olufsen was fishing off the coast of Bodø, when a monster with a human face and large claws suddenly grabbed hold of his boat when he tried to get his net in. Apparently, the creature had become entangled in the net. Olufsen was in a bit of a predicament, but probably lucky for him the net split, and the merman disappeared in the waves. Oh yes – there have also been sightings of smaller versions of merfolk – merkids, perhaps, locally known as *marmæler*. Some of them have even been seen playing, throwing stones and pieces of seaweed to one another.

Now what are we to make of all this? It is quite tempting to explain at least some of the sightings with garbled descriptions of seals of various species. One sighting off Bergen in 1898 described a merman with a red balloon on top of his head, which was almost certainly an actual sighting of a hooded seal, a bit outside it's normal territory. But not all cases can be explained this way, especially because there have been a few sightings of merfolk and seals together, and with a clear distinction between the two. In one case a dead merman and a dead seal were allegedly found on the coast near the city of Bergen. Both were covered in bite marks, and it was suggested that they had killed each other in a fight.

This kind of animosity is not unknown. One of my correspondents, a retired school-teacher living in Trondheim, has told me how it was common knowledge when he was a kid, or at least that's what his grandparents told him, that merfolk and seals hated each other, and would sometimes go into battle, fighting each other to the death in large flocks. You should also in general treat the merfolk with respect. In his grandfather's time it was still common in some parts of Norway to bring sacrifices to the merfolk on Christmas Day.

A curious collection of critters
Apart from the more "classic" monster-types, the seas around Norway is also the home of some very weird and wonderful ones – a small collection of which I will present without further ado…

There is, for example, the sea-ram. Now with a name like this, one would expect a creature generously endowed in the horn department and capable of smashing gaping holes in the sides of ships when sufficiently enraged. But alas no. This creature subdues ships and sailors by its very powerful and stinking breath, apparently something which can happen without warning or provocation. One minute you are sailing along, minding your own business and doing a spot of fishing, and the next this cloud erupts from the sea, reeking of dead fish and other malodorous miscellanea. If you are quick, and not completely overcome by this extreme case of bad breath, you might be able to see a big black creature with prominent white horns on the side of the head, speeding away just below the surface, probably chuckling to itself.

This sounds very strange, but although I have talked to a couple of people who have experienced a sea-ram first hand, and who swear that it was completely unlike anything they have ever come across before, I think the explanation is fairly straightforward. Anyone who has been in close contact with a whale, or more specifically their breath, can testify that they could do with a more regular use of mouthwash. Especially whales that live on small fish, as they can have eye-wateringly bad breath. So – a whale. As for the horns, well, the killer whale or orca has a rather prominent set of white markings on the side of the head. They vary quite a lot in size and shape, but are distinctly horn-shaped in many animals. And killer whales can have rather bad breath as well. Or perhaps we can envision a scenario somewhere along these lines – a larger whale, perhaps a minke whale, is being chased by killer whales. It has dived deep to try and avoid them, but eventually comes up next to a boat, exhales something stale, smelly and nasty, takes a deep breath and dives again. The fisherman, having picked himself up from the deck whereto the smell has knocked him, is too late to see the bigger whale, when he cautiously looks over the side, all he sees is the black and blurry outline of a killer whale giving chase, its hornlike markings clearly visible.

Then of course there is the mer-sow (we seem to have a whole range of domestic sea-creatures here – perhaps escapees from the stables of the merfolk?). The mer-sow is a serious creature, some 20 metres long, 5 metres deep and 2 metres wide, with a head like a pig, and no less than six eyes, three on each side. The colour varies a bit apparently, from pinkish grey to light brown, and to top it all off, it also has a big dorsal, crescent shaped fin.

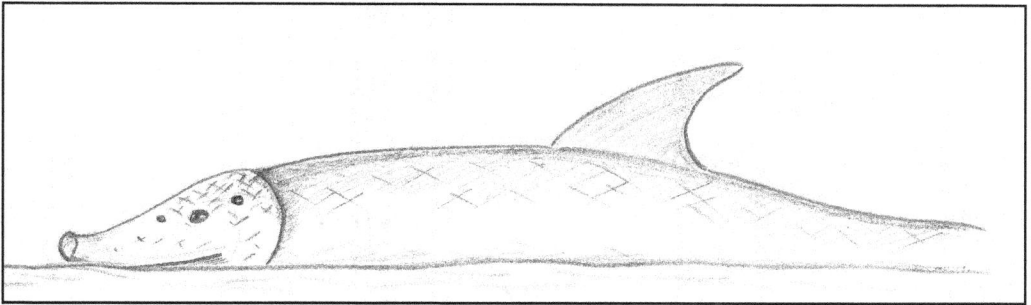

The author's drawing of a Norwegian mer-sow. Sounds strange, but if one stretches the pig-like head a little bit, we end up with something rather sturgeon-like. But three eyes on each side of the head...?

If the mer-sow had only been seen dead, I would not have hesitated to write it off as a badly decomposed whale or shark, or perhaps a walrus, but some people claim to have seen these creatures alive and kicking (or splashing as it were), and one has supposedly even been caught sometime in the 18th Century. Living mer-sows are apparently rather slow and ponderous creatures, and this makes me wonder if they might be garbled sightings of live walruses. I know they are not as big as that – and usually have tusks, but a big tuskless individual? And the colour fits quite well. Walruses do blush! Oh – and the three eyes? Well, perhaps a nostril, an eye and an ear-opening on each side of the head?

Another, but far more unlikely candidate - although not entirely impossible, giving the fact that mer-sows have not been seen for some 250 years - is Steller's sea-cow. I know they officially existed only in the north-western Pacific, but there are sightings here and there from other parts of the north Atlantic and the north Pacific of creatures that might be this giant sea-creature. In this case, the size of the mer-sow is a better fit. And the head of a sea-cow can, with a little bit of artistic license, be said to be pig-like. At least the contemporary species can. I have only seen them in Belize in Central America, and presumably the Steller's sea-cow would look similar.

And then there is the worm, orm, wurm or whatever you prefer to call it. I have spoken to a number of people about it, and they all just call it "*ormen*" – the worm. This rather strange creature is apparently an extremely long, 20 metres or more, worm. It is very thin, and at its biggest no more than the arm of a small child. It is harmless, unless you squash it, or step on it, or something, and then it will swell up to many times its original size.

Now that sounds like pure imagination one would think, but as a matter of fact there is a real creature even more strange than this: the bootlace worm, basking in the scientific name *Lineus longissimus* which can be translated into something like "the longest line". This creature is no thicker than a shoelace, or indeed a bootlace, but can be rather elongated. An individual that got washed ashore on a Scottish coast sometime in the 1800s was something like 80 metres long – that's three good sized blue whales in length! But it is far to thin to be the worm, so perhaps there is another species out there, not quite as long, but more bulky than "the longest line"?

Mus marinus, and two others too strange to be true?

In the Scandinavian languages some marine animals are named mice of some kind or other. There is the "*sømus*" – the sea-mouse - which is in fact several species of irregular sea-urchins, the "*guldmus*", the golden mouse – which is in fact a large, fat marine worm with a dense covering of iridescent hairs on its back, and finally "*havmus*" - the mer-mouse – which is in fact several species of fish belonging to the order Chimaeriformes. These are all present and accounted for, but according to a man from the Lofoten Islands of northern Norway, there is another kind of marine mouse in the Norwegian seas. This one was as big as a medium-sized dog, rounded and with fins and big eyes. It normally lived in the sea, but it would come up to lay its eggs on land, and then disappear into the water, only to return 30 days later to dig up the young, and bring them back into the sea.

Now, I don't think I was being had, as I have heard one similar story in southern Norway, in a house some miles southwest of Oslo. But I haven't got the faintest idea as to what could have given rise to this strange story. There are of course quite a few marine animals that lays their eggs on land – sea turtles spring to mind, but big mice?

The only thing I can think of is some kind of distorted story about seals. Could one perhaps theorise that someone, someday had seen some seals on a sandbank somewhere, perhaps rolling around on the sand, scraping it up and throwing it about, and this man thinks: "Gee, those animals that look like big fat mice, must be digging for something. I know – they are digging holes for their eggs!" And then he goes home, and perhaps a month later, some other person from the same village comes back, and sees the same "mice" on the sandbank, but this time with young ones. "Oh – they have come back for their young!" And indeed, within a short time they all take to the sea.

Thin perhaps, but I am open to suggestions.

The bishop's brawny beast

Before we start the story of the bishop's beast, I might also mention a couple of beasties that I have, alas, only single stories about, but which are still frustratingly intriguing. I would dearly love to hear from anybody that might know anything more about the *sahab*, a creature with a huge body, and a long extended foot that it uses to feed itself. It might be a squid, and then again, it might not. No matter, I would dearly love to hear more. And that goes for the *swam-fisk* (swam-fish) as well. This creature apparently covers itself with slime, thus appearing to be dead, luring all kinds of scavengers close, which it then promptly eats.

In the Year of Our Lord 1520, the Archbishop of Trondheim, Erik Walkendorf, sent a letter to His Holiness Pope Leo X in Rome. Accompanying this letter was a package with the dried head of a strange creature that had been seen off the coast of Norway, and of which one now had been caught. And the Archbishop thought it would interest His Holiness to see what strange wonders God had created in far flung corners of the globe. As the bishop so succinctly wrote, this creature looked like nothing anybody had ever seen before. It had a large squarish head, was black, and about 5-6 metres in length – not that monstrous one would say – and the body was only slightly larger than the head. A young sperm whale perhaps, or a dwarf sperm whale? That would be my guess at this point in the narrative, but then the bishop goes on to describe the very large red eyes of this creature. Right away we are in trouble, because whales actually have rather small eyes compared to their size. Then he clinches it by describing how this creature was also covered in long thick hair that almost looked like goose quills. Baleen somebody has suggested, but I find that very hard to believe. A question of decomposition then, as it is well-known things like, for example, sharks can develop what appears to be fur during the decomposition process. And then again, maybe not – after all, this creature was very much alive when it was first seen. I have no suggestions worth anything, but perhaps somebody with a better grasp of Italian than me could write to the Vatican some day, and ask them if somewhere in the files of Leo X there is a strange dried head.

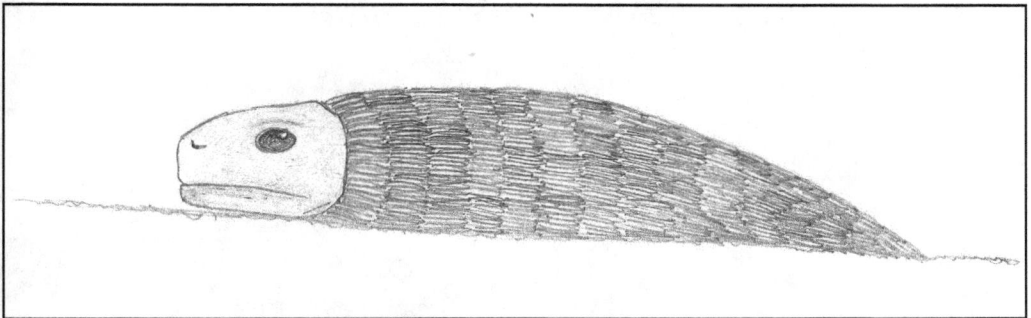

The descriptions of the Bishop's brawny beast are rather vague, but it must have looked something like this drawing by the author.

Horses with many skills

Now we come to a group of animals with a bit of an amphibious lifestyle – the merhorses. They have been seen in various incarnations, in the sea, on land, and in various lakes - the perfect introduction to lake-monsters as it were.

Merhorses and lakehorses or waterhorses, or perhaps I should call them horse-like aquatic beings, are a strange, and surprisingly wide spread bunch. You find them all over the place, not only in various north European countries, but also in Great Britain, various places in the rest of Europe, and even in America and a few places in Asia. I have even heard the odd story or two from Australia. Some of them are very horse-like, and some of them perhaps not so much.

If you are out to sea (in the most literal sense of the expression) and meet a merhorse, you will see a creature with a muzzle or a head like a horse, usually very big nostrils, big eyes – sometimes described as bright blue - with a dark brown or brownish skin, perhaps speckled, so far all very horse-like.   But with a pair of front limbs looking like flippers, and a long, fish or whale-like tail, and sometimes even two tails!

OPPOSITE: Sea-serpents are so common in Norwegian folklore that you can find them just about anywhere—in this case, a merhorse at the bottom of a fountain just behind the Town Hall in Oslo.

74

That is something straight out of a fairytale, but lots of people claim to have seen these creatures, although none of the sightings I have been able to unearth is recent. Some of them are quite interesting: in 1746 one of these animals was seen off the coast of Molde during a rather lengthy stay in the area. Lots of local people, fishermen and so on, managed to see the merhorse before it suddenly disappeared one day. As far as I know, nobody has ever managed to capture or kill one of these merhorses, although there is an old story of a rich man in Bergen having done just that, and having a tablecloth made of the skin. A tablecloth which was apparently much admired by the rest of the inhabitants in Bergen.

I do have one sighting from the outer rim of Oslo Bay, from sometime in the late 1880s. The observer was the great-grandfather of a man I talked to during a stay in Oslo. Apparently he had been fishing on a calm summer's day, when this seal-like creature - but with a long square snout - suddenly surfaced quite close to him, with a large fish in it's mouth, almost to mock the man, who hadn't been able to catch a single fish all day. The creature was apparently rather curious, because after having swallowed the fish, it swam closer, and subjected him and the boat to an intense scrutiny for several minutes. During this stay, the man noticed that the creature's large eyes were rather dark, but with a bluish hue, almost as if the animal was suffering from severe cataracts. He could also plainly see whiskers, and would have been content for the animal to have been a seal if it hadn't been for the long square muzzle, something you usually don't find in any seal. After a few minutes another boat approached, which made the animal dive under the surface and disappear.

I too would have been quite content, or actually am quite content for this animal to be a seal of some kind, although the exact species is a bit more difficult to ascertain. It was after all, according to my correspondent, some 6 metres in length, and there is still this strange squarish head to explain. The great-grandfather was quite familiar with the seals of the area, harbour- and grey seal, and he was adamant, that this was something entirely different. But what? I for one am at a bit of a loss. The leopard seal has a rather longish snout compared to other seals, but it is not 6 metres in length, and it lives around Antarctica. The elephant seal can be very big, although 6 metres is a bit much, but an elephant seal would have been a rather long way from home. A walrus is not that far off, but for the facial aspect of such an animal, as they usually have more of a porcine aspect, definitely not equine. And 6 metres would be one hell of a walrus!

So what are we to make of this? If it were a one off sighting, it would not be that difficult to invoke perhaps a malformed seal scaring the poor man out of his wits, as well as his ability to judge the creature's size. But quite a lot of these beasts have been seen. Some of them even on land!

Horse-like creatures that are capable of running around on dry land are quite common in northern Europe. They are known from Great Britain as well: waterhorses, kelpies or whatever local name they have. In all cases they are sufficiently horse-like to fool a lot of people, although if you look a bit closer, there is something very strange about them. They graze like any normal horse would do, but they have cloven hoofs like a cow. They are capable of elongation and shape-shifting too. Some of them prefer to escape into the sea if being pursued by humans, others will run in the other direction, and dive into any suitable lake – and this brings us to the subject of lake-monsters.

Selma

By far the most famous of the Norwegian lake-monsters is "Selma", the creature from Lake Seljordsvannet in southern Norway, not the least because it has been the subject of several expeditions arranged by the so-called GUST-organisation. Much has been written about Selma; some good, some bad, some bordering on the ridiculous, and untangling all of that would probably take a book in itself, so I will make do with a rather short summary, as there are several other lake-monsters worth mentioning, and we haven't got thousands of pages.

Most books about lake-monsters will tell you that the first described sighting of Selma is from 1750, when a strange horse-like creature was seen swimming in the lake (those waterhorses again!). But the stories probably go back even further. Some years ago, I had a lengthy talk with Olav who was 86 years old at the time, and had been a priest in one of the little towns close to Selma's lake for most of his life, so he was intimately familiar with the stories. He was also a bit of an amateur historian, and according to him the stories went way back, probably even to medieval times. He had found a short note about a sighting from 1645 of a "waterhorse" in the lake – intriguing, but unfortunately, there was no more information, and another one from 1612, which indicated that the knowledge of the creature was well rooted even then.

The sightings from the lake are incredibly diverse and varied, and it can be a bit difficult to make head or tails (tales?) of them. You can meet the classic "upturned boat" as well as a snake-like creature with a horse-like head. But there are also a few that are a bit more strange. In 1880, some local claimed to have killed a strange 1 metre long animal with four legs (a youngster?) and very large eyes. The whereabouts of the remains of this beast was never ascertained with any degree of certainty, so what it actually was nobody really knows. But perhaps this same creature, or one just like it, was seen in 1920, when a fisherman claimed to have seen a crocodile-like beast coming out of the lake and going ashore.

There have been a number of sightings of various beasts in the lake since then. The latest I know of is from the summer of 2009, but nothing has unfortunately been caught or photographed, so we are nowhere nearer understanding the nature of Selma than we are the Loch Ness monster, to name but one.

Regarding the crocodile-like being of the 1920s, there might be some kind of explanation, though. The creature was apparently seen quite a number of times in the summer of 1920, and according to some local people I have talked to, and whose great-grandparents lived around the lake at the time, the creature not only looked like a crocodile, it was a crocodile. It was common local knowledge, and most people claimed to know who had released the beast into the lake, although they wouldn't tell me.

However, fortunately there is more to Norwegian lake-monsters than just Seljordsormen. Quite a number of lakes sport animals and sightings of the monstrous kind.

The others...and then some!
The story of many of these creatures goes back a very long way indeed. Some sightings are from the 14th or 15th Century, and in some cases even older.

Lake Mjøsa, which - with a length of 117 kilometres and a depth of up to 468 metres - is not only the biggest lake in Norway, but one of the deepest in Europe, is the home of one of the oldest monsters. The creature has been seen at least since some time in the 14th Century. Allegedly the first sighting was when one of the animals was killed in the lake for some reason, and the bones, which were unknown to anybody who saw them, later washed up on the shore. Unfortunately, nobody seems to know what happened to the bones. I found in one description that they were buried, whereas another one claimed that they were thrown back into the lake again.

The monster, which has been seen several times since then, is described a having a horse-like head, a black mane and very large eyes. The number of sightings is not that great, and the newest one is from the summer of 2005, although there are probably more if you really start digging. The observation from 2005, done on a very hot day in July, describes a long snake-like creature with a horse-like head – very much true to form. A similar creature, affectionately known as "Rømmie", can be found in Lake Rømsjøen. It is remarkably similar to the creature in Lake Mjøsa: dark, with a mane and a snake- or log-like body, although the head of this one is described a being more calf-like than horse-like. This one has been known since the 18th Century, but the legend was already well established at the time so the creature is probably far older than that.

A school bus driver and 15 kids he had in his bus for example, saw it on 20th September 1976. They saw it swimming across the lake with it's head held high, clearly turning from side to side looking around.

Some years ago – in 2001 to be exact – I managed to track down two of the kids on the bus (although they hardly qualified as kids anymore) – one of whom flatly refused to talk to me. The other one reluctantly agreed to tell me what she saw, if I guaranteed never to mention her name in writing, explaining how she had been teased mercilessly for years about her sighting, and at one stage even had to change jobs because of it. She told me that she had no doubt that she had seen a live animal. The body was the shape of a large log, but undulated gently from side to side while the creature, which was swimming away from them, was constantly turning its head from side to side. The colour was uniformly dark, and the body seemed very slick and shining with no kind of surface structure.

The animal has even been seen on land on at least two occasions that I have been able to track down. In one case in 1965, it was a blistering hot summer's day, and it was on its way down towards the lake, apparently sliding on its stomach like an otter playing.

On the other occasion the animal was seen late one evening in 2004, moving across a road and disappearing into the water with a motion the witness described as something similar to how a sidewinder snake moves in a desert of hot loose sand. The size of the various lake-monsters apparently varies quite a lot, from 2-3 metres, all the way up to something in the vicinity of 20 metres. One of the very few aberrant lake monsters can be found in Lake Myrkevatten. This one is green rather than blackish, and sports a prominent beard every time it rears its head and neck up from the surface of the lake.

Kill it!

One of the things I am often asked when talking about lake-monsters, is why nobody has caught or killed one of these creatures, and I can only put it down to bad luck. It is not for lack of trying – believe me. As early as the 16th Century a lake-monster was killed in Lake Reins-vannet, and was dragged ashore. Unfortunately, nobody cared to preserve any part of it, and it was buried under a large mound of stone, probably to keep it from haunting the place. Today nobody is quite sure where the burial took place. I for one would jump at the opportunity of a bit of excavating.

Then we have the monster in Lake Flåvannet. In 1880, the steamer plying the local waters collided with the creature and killed it. There was plenty of blood and oil in the water, but it sank out of sight before anyone could do anything about it. There are also a number of hunters who have taken pot shots at these creatures, but all to no avail. Either the animals seem singularly unaffected (probably because the shooters missed) or they disappear under the surface, never to be seen again. And who can blame them?

Whodunnit?

Lots of sightings, lots of stories, lots of different creatures. But when all is said and done (and written), the one burning question still remains. What are they? What is going on?

Many theories have been put forward by way of explanation, some of them bordering on the ridiculous (secret government experiments with trained dolphins or mini-subs spring to mind), some of them far more reasonable, but none of them can explain every sighting, probably because we are dealing with a whole range of phenomena, all grouped together under one heading – "lake-monster", and only some of these have a zoological explanation.

My personal favourite, something I have presented as a way of explaining most of the sightings of the Swedish lake-monster Storsjöodjuret (see the chapter on Sweden for more details on this), is the swimming moose. These big ungainly animals are far better swimmers and divers than most people give them credit for, and they could easily explain the horse- and calf-like head. If there is something out of the ordinary living in these lakes it must be piscine - perhaps the great sterile eels that Richard Freeman and Jon Downes of the CFZ have advocated. They would certainly explain the snake-like way of moving, even on land, which is something eels are quite capable of doing, as well as the dark or blackish colour.

The rest - as Shakespeare would probably have put it - is not silence, but a dash, a fright, a sprinkle of imagination and some good old-fashioned misinterpretation of wakes and other well-known phenomena. But I still have no idea what the green monster in Lake Myrkevatten is.

SWEDEN

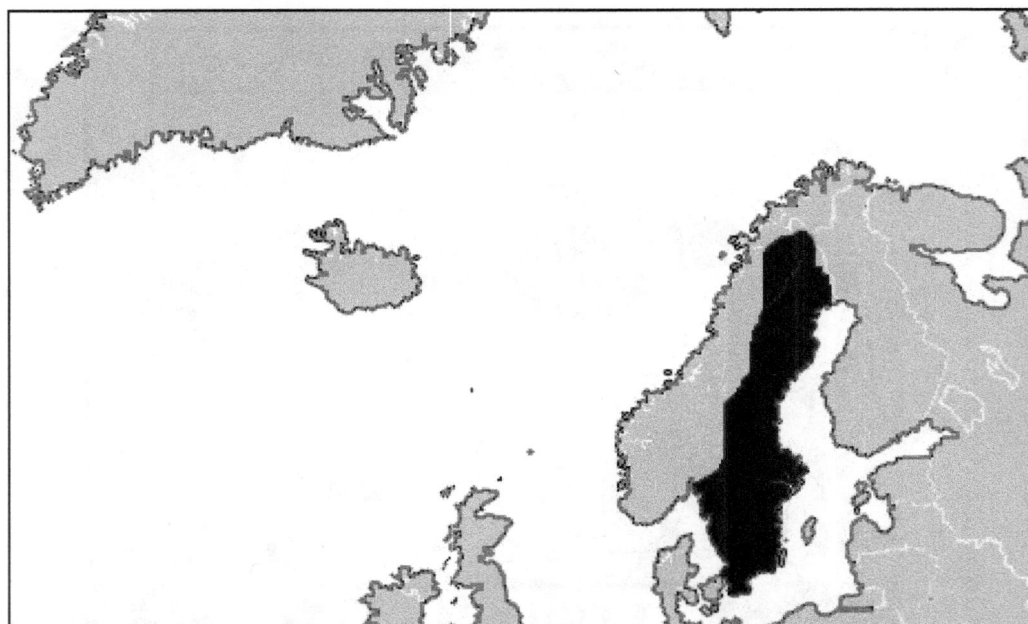

Sweden is probably the only European country with a lake-monster whose fame comes anywhere near the Loch Ness monster. Storsjöodjuret – a long and rather daunting Swedish word – can be translated quite easily: "Stor" means big or great, "sjö" is the Swedish word for lake – thus Great Lake, and "odjuret" is the Swedish word for the monster, so all in all we end up with The Great Lake Monster. This interesting creature has been a stable part of lake-monster books for decades – the present one being no exception. Nothing wrong with that of course; the great interest in the creature has ensured that we know quite a lot about it, and it has a vast number of sightings, but at the same time it is also a problem, as Sweden's other lake-monsters tend to be completely overshadowed by their more famous relative. That is a shame, as some of Sweden's other aquatic monsters are rather interesting, and rather strange.

Correction ...

...Of sorts. In various Nordic legends, and in various books of same, as well as some cryptozoological tomes, you will read descriptions of rune-stones with depictions of various lake-monsters with a famous example being situated on a hill overlooking Storsjön. But take no notice. These animals are not lake-monsters. Those legends have grown up around them in the centuries after the Vikings erected them. They are in fact depictions of Jormungandr, the Midgårdsorm, the vast sea-monster that, according to Viking lore, encircled the world and was so enormous that it could bite its own tail.

Ladies and gentlemen! I give you....Big Ears!

Much has been written already about the monster in Lake Storsjön, and I will not try to repeat it here, apart from the fact that there have been lots of documented sightings since the end of the 1800s, and probably lots more before that. There have also been various attempts to catch the beast, and you can still see the traps and various other contraptions used if you visit the museum in Östersund, the biggest town by the lake. At one stage, the local authorities even gave the animal protected status.

However, the main thing is that the Storsjö-monster is very similar in many ways to other northern hemisphere lake-monsters, apart from one crucial difference. This mighty Swede is blessed with a set of very large whitish or light brown ears with serrated edges, making them look somewhat like a set of bat's wings. Oh yes, and the monster also has a habit of suddenly appearing from the depth of the lake next to small boats, and scaring the pants off anyone close by. Apart from that there are heaps of sightings of mysterious wakes on the lake, the back of something big swimming just below the surface, and a few sightings of a neck adorned with a large horse-like head.

This runestone near Lake Storsjön allegedly depicts the monster - in actual fact it is Jor-mundgandr, the Midgard Serpent.

All this has been known for a number of years, but for some reason nobody seems to have made the connection between a lake-monster with very large "ears" in an area with a very large and healthy population of moose. I am not saying that there is no unknown animal in Lake Storsjön, but I am quite convinced that the vast majority of sightings can be explained by sightings of swimming moose.

When you see a moose on land you are immediately struck by its large and rather ungainly figure. It looks like it hasn't quite mastered how to control all of its legs at the same time. But do not let that look deceive you. A moose can move with more grace and elegance than a stealth bomber. Despite weighing in at something like half a ton, a moose can move incredibly quietly. I have been in many Scandinavian forests, and have been shocked out of my wits several times when suddenly finding myself face to face with a moose, after having not heard a single leaf rustle or a branch break. And if this isn't quite enough to rattle your nerves, moose are excellent swimmers, even better than people who *are* used to them give them credit for. They can swim long distances, and every now and then a moose swims from Sweden to Denmark and lives quite happily in Denmark, usually until they are knocked down by a car or a train. They can also dive down to around 4-5 metres for aquatic plants, something of which

OPPOSITE: At the local museum near Lake Storsjön it is still possible to see some of the traps local people have used in their attempts to catch the monster.

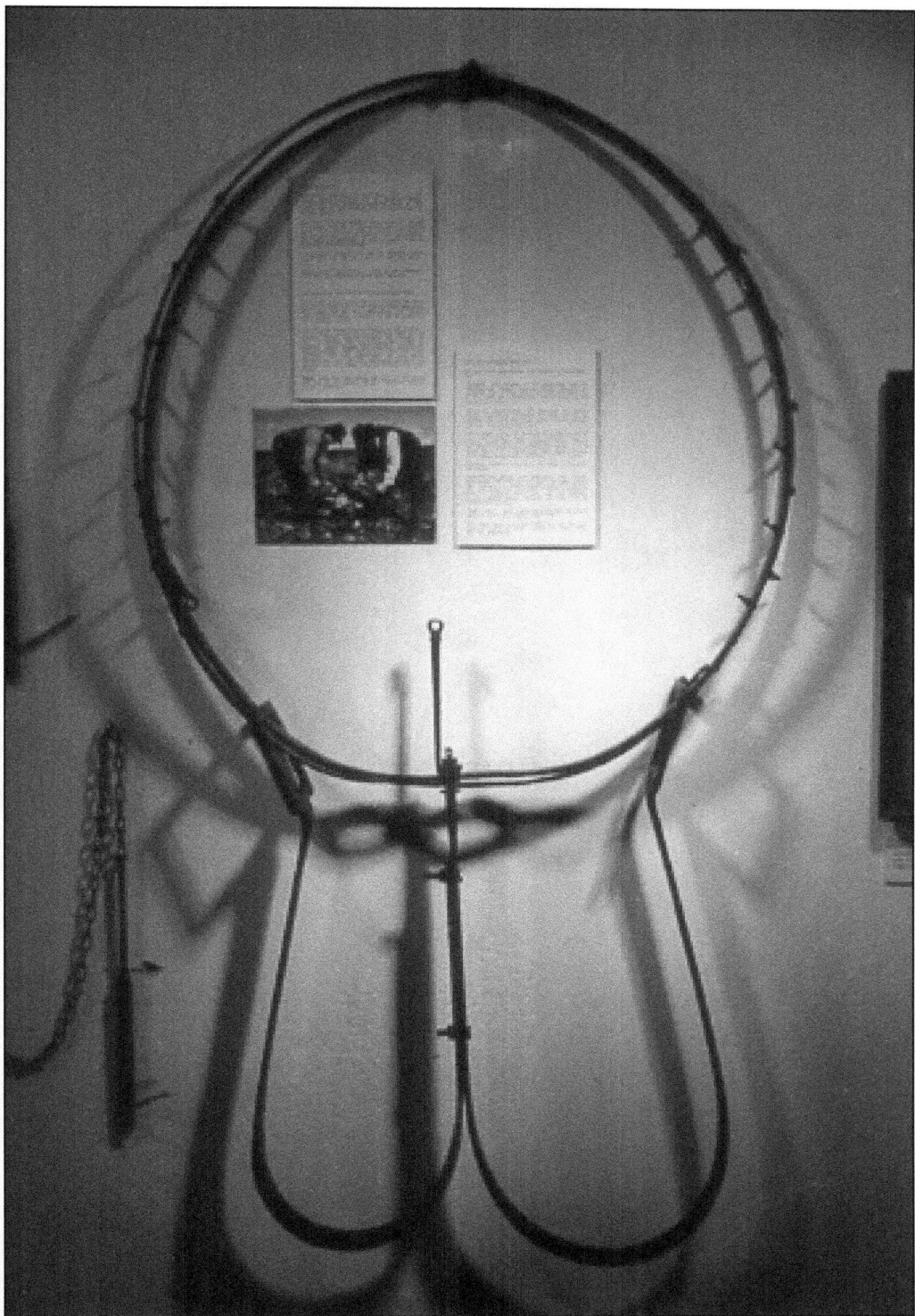

The strange `ears` on the head of the storsjöodjuret monster are probably nothing more than the antlers of a moose

they are extremely fond. And then of course they are, at least for part of the year, in possession of a very impressive set of antlers, looking for all the world like a big serrated pair of bat's wings or something similar.

Correct me if I am wrong, but imagine you are sitting in a small boat - perhaps doing a spot of fishing or just thinking nothing - when suddenly, next to your boat, up comes this enormous dark brown creature, with a set of clothes-hangers like you have never seen before, and shakes its mighty horse-like head, perhaps with a big mouthful of water plants. Would you not think you had seen a monster? Or perhaps wet yourself. Or perhaps both? I probably would!

A moose paddling around looking for aquatic plants, perhaps with its head under the water, but with its back above, would also do nicely for an "upturned boat" monster. In this case, I don't care how much people think they know moose, this is unlike anything you have ever seen.

OPPOSITE: The symbol of the city of Östersund - right on the banks of Lake Storsjön

There has been the odd sighting of something much lower in the water swimming just below or exactly at the surface. There are even a few photographs. Some of these do not look like swimming moose, and some have suggested that these sightings are evidence for the existence of wels catfish in the lake. I seriously doubt that since Storsjön is several hundred kilometres north of the normal distribution of this huge fish.

But should you get to Storsjön, do take your time. It is a delightful place, and a good time can be had by all.

The other lakes

Apart from Storsjön, there are a couple of other very big lakes in Sweden, one of them being Lake Vättern where fishermen for centuries have complained of their nets being torn, and have reported sightings of a humped animal sometimes being seen on the surface. In 1947, a local tugboat called *Hebe* even reported a sighting of no less than three of these animals, each one about 15 metres in length and with two humps each. As late as 2001 and 2007 there have been other sightings of humps, but nothing newer.

So far, this very much looks and sounds like traditional lake-monster lore, but the lake also contains something far more along the lines of *Creature from the Black Lagoon*, although to

the best of my knowledge, this beast has not been seen for more than a hundred years.
This thing, last seen in 1897, where it walked out of the water on two (!) legs, was described as being man-like, extremely ugly, and covered in scales or perhaps barbels and fringes, tufts and various other form of outgrowths. I have absolutely no idea what this is all about, but apparently the creature's existence was well known in the area, so it must have been seen quite a number of times.

What is it with lake monsters and priests?
Another large Swedish lake with a long and colourful history of monsters is Lake Mälaren, just west of the Swedish capital of Stockholm. The sightings are very similar to what have been seen in other lakes, so I will not go into details here, but only describe the rather strange religious twist on things here – shades of St. Columba and the Loch Ness monster and several other similar encounters between priests and monsters (the priest always wins for some reason).

In 1652 French priest Pierre-Daniel Huet, later to become Archbishop of Arranders, visited Sweden and was told not to swim in the lake because of the monster. Unfortunately I have never been able to find any description of whether he actually tried, although I am quite certain he would emerge unscathed – they always seem to do that, don't they? It would have been fun if he had been attacked and had gotten a chance to flex his religious muscles.

The lakes of Skåne
Most Swedish lake-monsters have a tendency to keep themselves to more northerly lakes, but even in the extreme southern part of the country there is a wealth of monsters to be seen and studied, but with a fair bit of luck as they do not show themselves often. However, the various lakes in Skåne are popular tourist attractions and stops for day-trips, as well as for bird-watchers, of which there is a considerable plenitude in Sweden as well as Denmark.

Börringe, Sövde, Vomb and Kranke are names well-known to anyone in Scandinavia with ornithological interests. They are a few of the many lakes in this area between Malmö and Ystad, the two biggest cities in southern Sweden. All of them are known as great places for bird-watching, especially if you are interested in heavy-duty birds like golden eagles and white-tailed sea-eagles.

Something that is not equally well-known is the fact that the various lakes are also home to a varied fauna of monsters in no way inferior to the more well-known examples of this mysterious group of animals, such as the Loch Ness monster from the Scottish Highlands.

On a global level you can roughly divide the sightings of lake monsters into four groups:

1. Disturbances and splashing in the water, traces of a wake with no boats in sight and similar aquatic phenomena, which might conceivably be produced by something large moving about beneath the surface.
2. Something looking like an upturned boat drifting around. If there is a current or wind, it is very common for the "boat" to move against the current or the wind.

3.   A long, thin neck with a small head rising from the water. In rare cases you also see a body, often very powerful and elephant-like.

4.   An elongated snake- or eel-like creature often measuring 5-6 metres or more.

All of these types have been seen at some time or other in the lakes of Skåne. Not all of them are equally common in all places, and you will not find all the types in any one lake.

There are so many lakes in Skåne, the stories would probably fill a book in themselves, but I have chosen the four lakes mentioned because I am thoroughly familiar with them. Besides being a cryptozoologist, I have also been an eager birdwatcher for more than 30 years, and I have visited all four of the lakes numerous times. I have personally never seen anything mysterious in any of these lakes, but I am familiar with the terrain and know exactly how hard or how easy it is to see something.

The only time I have ever seen anything strange in the area was at the end of January 1999 in a lake called Krageholmsjön. I had made a stop with some friends looking for white-tailed sea eagles, which you sometimes see sitting on the frozen lakes looking like miniature haystacks with rather piercing eyes. On that particular day two eagles were sitting out there looking rather depressed, and while we were studying them, freezing our behinds off in the process, we suddenly heard a loud cracking sound. About 75 metres from the shore, roughly halfway between us and the eagles - which didn't seem to notice the cracking - the ice was suddenly pushed upwards from below. One of my friends suddenly exclaimed in mock fright: "My God – it's a submarine!" But we never did see the conning tower or anything. We did see the ice lift, as if something was pushing it, but we could not see the 'something'. Nothing further happened, and the broken ice stayed up, but I still wonder? Just movement of the water? I do not think the ice was actually breaking as it was 10 degrees below zero. What was it? I wish I knew.

Anyway – back to the four lakes:

- **Lake Börringe, October 1988:**

Eight members of a night school class from Copenhagen are on a one day trip to Skåne looking for eagles. There are all in the middle of eating their lunch at the shores of the lake, when one of them spots an area of severe disturbance several hundred metres out on the lake. It is a very calm day, and apart from a few ripples on the surface, the lake has been calm until then. Though all eight have binoculars, none of them are capable of discerning what is going on in the lake, but several of them think that a couple of times they get a glimpse of one or several large animals splashing about just under the surface and churning up the water.

I heard the story a couple of years later, when I met two of the group at a site further east in Skåne, where we were watching no less than eight sea-eagles doing their haystack impressions. They still had no idea what they had seen, but later that day they showed me the exact spot of the sighting, a promontory almost exactly in the middle of the east bank of the lake. At that particular point you are almost at surface-level, so it can be difficult to see something in the distance, but I have no doubt as to their veracity.

- **Lake Kranke, May 1979:**
Very early one morning, two people standing in the old observation tower at Silvåkra on the eastern side of the lake discover what they first think is an upturned boat (drum roll, please!) drifting far away on the lake. At first they give it scant attention as there are a lot of boats on the lake, and they assume it is one of the local's craft that has slipped its moorings, and is now on the loose. There had been a severe storm a couple of days before their visit. But suddenly they realise that the boat is not drifting slowly about, but actually sailing back and forth on the deepest part of the lake. They discuss the sighting for a long time, but abandon the talk when the "boat" suddenly disappears under the water. "It was almost like seeing a submarine diving", as one of the eyewitnesses later put it.

- **Lake Vomb, November 1991:**
An elderly couple on a Sunday drive have stopped at a picnic site on the south side of the lake to enjoy the view. The lake is fairly shallow, but the biggest in the area.

Eyewitness drawing of the monster in Lake Vomb in southern Sweden.
Notice the rather prominent eye.

There is a slight mist on the lake, and the picnic site is some 10-20 metres above the surface, but the couple has no trouble watching the action when a large animal suddenly pokes a long neck and a little head out of the water, almost directly below where they are standing. The animal looks around two or three times, and they can clearly see its rather large eyes, before it dives under the water and disappears again. They wait for another half hour, but the animal doesn't return.

- **Lake Sövde, July 2005:**
It is a very hot summer's day, when a family on a picnic trip decides to take a break at the edge of a little forest, where they have a nice view of the lake. After about 15 minutes they are just about to leave, when the youngest member of the family, a girl of 8, discovers what she calls a big snake in the lake. None of the family members has brought binoculars, and since the animal is at least 300-400 metres away, it is very difficult to see any details, but the father later describes is as something like an enormous eel swimming at the surface. He judged it to be at least 10-12 metres long.

There is an interesting aside to this. Stories about giant eels are quite common in Sweden (there are even a few from Denmark), and you find them in various literature and movies as well. In the Swedish TV-series "Bombi Bitt och jag" (from 1968), based on a very famous Swedish novel, there is even a fight between a boy and one of these eels. One cannot help but wonder if these giant, perhaps sterile eels – the existence of which has been suggested by Jon Downes and Richard Freeman of the CFZ on several occasions - had in fact at one time been far more common and well-known?

There are a few other sightings that suggest these giant eels are not that rare. In 1930, a baby "worm" was seen crossing a road near Bocksjön, another Swedish lake. One sighting described is as moving with caterpillar-like movements, which I presume means a kind of vertical undulation, but other sightings said it moved like a snake, which would be more consistent with an eel.

Various other lake-monsters have been described as snake-like, for instance the monster in Lake Gryttjen, which was very active in the 1980s. One eyewitness saw this creature, or one of them, swimming right next to a boat he was sitting in. He said it was a dark greyish/ blackish colour with a pointed snout, and with what looked like a couple of small fins just behind the head. Again, very much like an eel, although considerably bigger. He judged it to be something around 4 metres in length, and the thickness of an arm.

Salty tales
In spite of Sweden's very long coastline, the actual number of sea-monster sightings is fairly limited, perhaps because a substantial part of the coast is along the Baltic Sea, a rather brackish area, which may not be suitable for large sea-creatures. Perhaps also because large parts of it are frozen in winter. Whales are, for example, rare in these waters, as are even porpoises and such like, and the number of seals is also quite low. But strange creatures have been seen, and *are* being seen.

The merman of Ven
A very early sighting dates from the year 1723, when a merman was seen several times in the waters around Ven, an island situated in the narrow sound, the Øresund that separates southern Sweden from eastern Denmark. The sea-monk described in the chapter on Denmark, was caught in that same general area. The merman of 1723 was described as being large and very heavily built, perhaps even fat or at least severely obese, with big rolls of fat all over his body. He had rather small piggy eyes, and a severe beard of some kind. He was apparently relatively unafraid and curious, and often came rather close to small boats and fishermen, usually scaring the poor people out of their wits, or their pants (which sometimes needed cleaning afterwards). The merman had short stubby arms with big hands that looked like flippers.

Now I am wondering if this creature, which was described as having a brownish tan, but was also capable of blushing, was not a walrus severely off it has beaten track. It all fits rather nicely together, except one would imagine some degree of knowledge of this animal even in the 1700s.

Tales from the bridge – and before

On the opposite side of Sweden, Kalmarsund, the narrow stretch of water separating the east coast from the island of Öland, is very similar to Øresund, and indeed there have been a number of sightings of sea-monsters here, although of a kind very different from the above mentioned merman. The animals seen here are very large, with a body like a big tree-trunk. To my knowledge, it has never been seen on land, or even above water, but quite a number of people claim to have seen it swimming passed just below the surface. It is usually described as being 5-6 metres in length, with no discernible limbs apart from four short stubby flippers.

Motorists driving across the bridge now connecting Öland with the Swedish mainland report a number of modern sightings, but there were also sightings from before the building of the bridge. It is a bit difficult to suggest what we are dealing with here. Seals, porpoises or perhaps dolphins are a possibility, but being mammals they should at least sometimes come to the surface to breathe. Or maybe it is minisubs? Confused? Read on, all will be explained, for now we come to the area around the Swedish capital of Stockholm, the so-called "skærgård", consisting of hundreds of big and small islands. Some very strange goings-on have been reported in this area. Here we have to look at a mix where the Cold War, paranoia and politics play a part as well.

Saltie

It all started in the early decades of the 1900s. From about 1900 until 1920, the skærgård was the location for numerous sightings of a creature affectionately knows as Saltie. All cases described it as a 20-30 metre long snake with the head of a turtle. One of the last sightings in 1920 was of no less than two animals seen simultaneously. As far as I know, no photographs were ever taken of this creature (probably a bit early for that). If somebody out there can prove

Eyewitness drawing of the head of Saltie - somewhat turtle-like, and with a touch of cartoon animal thrown in as well.

me wrong, please do! But I do have in my possession a drawing made in the early 1970s by a man who had seen the creature in 1918, when he was 8 years old. The drawing was made many years later, it has to be said, and was given to me by this person's younger brother, who at the time was a colleague of my father (a musician playing in the symphony orchestra in Malmö in southern Sweden). It only shows the head, and it is indeed very turtle-like.

The reason for this could very well be that it was in fact a sea turtle. I know sea turtles are not an ordinary element of the fauna in northern Europe, but every now and then individuals appear to get severely lost, and end up in Danish waters as well as Swedish. So why not a leatherback turtle, or perhaps several leatherback turtles showing up around Stockholm? A bit of a coincidence, but sometimes things do happen in almost epidemic-like proportions, like the giant squids that suddenly stranded almost by the dozen on the coasts of Newfoundland in the 1890s. I have no solid evidence of course, but it is a thought worth pondering.

Apparently, the scientists at the museum in Stockholm were not really interested in the creature so nobody tried to make a formal scientific inquiry into the matter. In 1920, the sightings stopped as suddenly as they started, and nothing was seen or heard from Saltie again. And for some time, quite some time in fact, all parts of the skærgård were at peace.

Subs
But then in the 1980s, things started to pick up speed again, and strange sightings were made all over the place, only this time people were not seeing sea-monsters, they were seeing submarine periscopes. Sea-monsters they might not be, but they give an idea of what strange things *can* be seen.

It all started in 1981, when a genuine proper Russian submarine was stranded near Karlskrona a little bit south of Stockholm. This happening, in the middle of a very paranoid period of the Cold War, released a flurry of sightings of periscopes and sub-marine hunts, depth charges being fired and sundry other exciting happenings. There were even sightings of something looking like caterpillar tracks on the seabed in various locations. Some of the sightings actually indicated living creatures instead of submarines, but untangling them could easily end up being a book in itself. That will have to be left to another time, and perhaps another author.

THE LANGUAGE BARRIER

Researching folklore and cryptozoology is fairly easy, if you do it in your home country, or in a country where the people and you have a common language. But when you get to the locally written sources you can easily end up having problems. Folkloristic subjects are often only of local interest, so they are published in the local language. Fine if you can speak or read the language, but if you cannot, or cannot find one who does, and who is willing to translate, you hit an insurmountable language barrier, which basically means that you can be denied access to a wealth of material. Quite a lot of countries are severely under-represented in cryptozoological literature. Some may think that this is due to these countries having no strange creatures, and thus no stories about them or eyewitness accounts, but the real reason is probably the language barrier.

And the language barrier very much comes into play when we reach the final four countries in this book: Finland and the three Baltic countries Estonia, Latvia and Lithuania. They do not have much to brag about when it comes to seacoasts, and consequently only a very few stories about sea-monsters, but all four countries have lots and lots of lakes and wetland, and therefore should have plenty of lake-monsters. Finland alone probably has more lakes (around 35,000 of them) than all the rest of the countries put together. Most people in Finland speak Swedish, and some of the literature is written in Swedish, so being from Denmark, I have no trouble with that, but proper Finnish, as well as the three Baltic languages is beyond me. So the amount of material in this book, although probably more than has ever been published in English before, does no real justice to what is actually there.

FINLAND

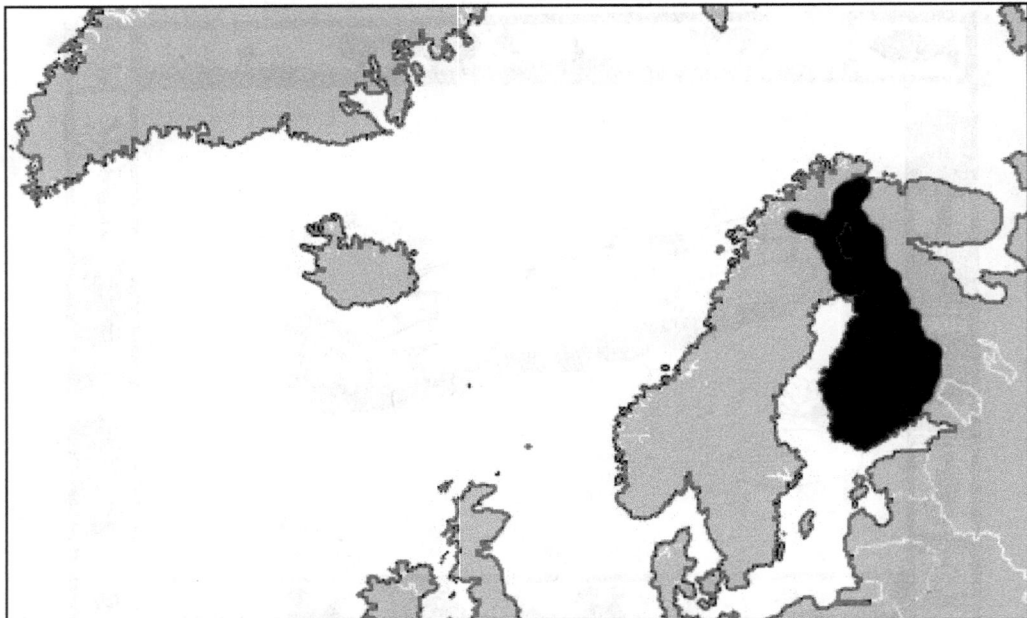

I f ever there was a country where the possibilities of lake-monsters were great – it must be Finland. Locally known as "land of the thousand lakes", there are actually at least 35,000 according to the most minimalist counts. It has more lakes than most other countries in the world, and they vary in size from the small and shallow, to the very large, and in some cases, quite deep. But the actual number of known sightings is surprisingly small. This may in fact be due to the aforementioned language barrier – Finnish is a fiendishly difficult language to learn (if you haven't grown up with it) and though a lot of Finns speak Swedish, which makes it a lot easier for a Dane to understand what is going on, most of the folklore is only written in Finnish. Another obstacle is the fact that the Finns in general, at least the ones I have talked to, have demonstrated a strange reluctance to speak of the more mysterious members of their fauna. Not because they are ashamed of them in any way, but more because they seemed to feel that is was none of my business. Maybe they were afraid I would report things in a misleading way. I know from my other line of work, that of a freelance writer of popular science, that the Finns like to their facts kept straight. Every time one of my pieces is translated into Finnish, it takes at least two eternities because they check and recheck everything, and then – just to make sure – they check it all over again. Perhaps lake-monsters are a bit to disorderly for their tastes, and so they will rather not talk about them.

Nevertheless, a few sightings have been made, and a few stories have been told, though in some cases it has taken quite a bit of persuasion.

The swimming logs

The Finns are a proud and somewhat melancholic people with a unique cultural style and sense of creativity. Finland has been the birthplace of some of the world's most unique and interesting composers and architects to name but a few. Therefore, it seems quite fitting that they should have some rather unique lake-monsters as well.

Several of the country's larger lakes are apparently home to a beast locally known as "the swimming log". This creature, some 3-4 metres in length and with a body like a big gnarled tree trunk, is usually seen swimming or drifting slowly about on the surface. One could probably argue that this is just the Finnish version of the upturned boat we know from so many other lakes around the world. In some cases, this might very well be so, but not all can be explained so "easily". Some of the Finnish "logs" are white, and what is even more interesting is that every now and then one of these creatures rears its ugly head above the surface. And it is not a small head on a long thin neck, but, on the contrary, it is a big, triangular head looking for all the world like a gigantic salmon.

This rather suggests that we are dealing with some kind of fish, although it must be of a species hitherto unknown in these parts of the world. This is corroborated by a bunch of stories I have collected about fishermen hooking something very large that simply just pulls the rods from their hands, or basically just swims away, dragging whatever boat the fishermen happen to be sitting in for several hundred metres before the line snaps.

An old Finnish gentleman I once met at a dinner party in Stockholm told me that when he was a boy, he had actually tried this once. In 1945 a swimming log was seen in a lake called Längelmävesi, and shortly after that, he went fishing in said lake. For several hours, absolutely nothing so much as nibbled his bait, but suddenly something just took off with it, not particularly fast, but slowly, steadily and completely unstoppable. "It felt like I had hooked a whale. Whatever it was, it swam off with all the line I had. And then the line just snapped. And not once did the animal come to the surface."

The only problem with this is, of course, that freshwater fishes in this size range usually keep to the tropics. There are a few possible candidates; sturgeons can grow to very large sizes, but they have a long pointed head that in no way resembles a salmon. The wels catfish is another potential giant, but apart from the fact that Finland would be quite a distance north of its usual range, the wels is not especially salmon-like. It has a big, broad and slightly flattened head, with a mouth the size of a barn-door in big individuals.

Members of the carp family can be quite large as well, and have a more normal "fish-shaped" head – perhaps salmon-like, if you stretch the issue a bit – but 3-4 metre long carp is a bit much.

Finally, there is also the possibility that the head of this creature looks like a salmon because they are in fact huge salmon – perhaps sterile individuals that have just kept growing and growing and growing.

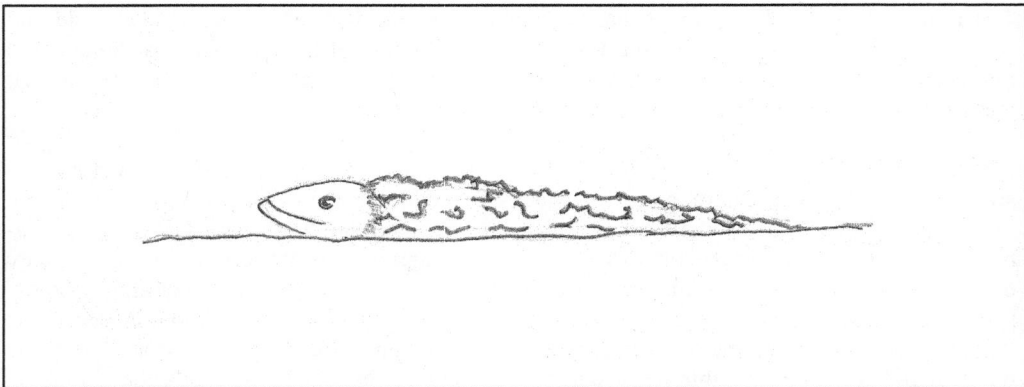

An eyewitness drawing of a swimming log from a sighting of 1988

The only problem I have with this is that carp and salmon have rather prominent back fins and big tail fins, and not a single one of the sightings I have collected mentions this. But I suppose you could put that down to the excitement of the moment.

King of fishes

Should these stories fail to satisfy your appetite for stories of the BIG ones, there is hope for you yet, for Finnish lakes are also home to a creature so fantastically huge, that stories of the legendary Kraken immediately spring to mind. These creatures, generally known as kings or queens of the fishes, are so big that people routinely mistakes them for islands, which I think is quite natural as they have grass and trees growing on their backs.

But should you be so unlucky – or stupid – to try and step ashore onto one of these islands, woe is upon you, as the beast, irritated by your presence, will suddenly dive under the surface, leaving you to die a horrible drowning death.

I have never been able to find an eyewitness to a meeting with one of these creatures, although that of course could be difficult if anyone that sets foot upon them drowns, but I have few stories that might indicate the actual identity of these monsters – something that could even explain some aspects of the Kraken stories.

For this we have to make a short detour to Fiordland National Park in New Zealand – a truly fantastic place. I have visited Fiordland twice, and on the second occasion I went sailing out through the very long and convoluted Milford Sound, looking for dolphins, penguins, albatrosses and other interesting wildlife. I saw a lot of good stuff, but at one stage I also saw a group of trees floating upright – with leaves and all. It was a truly strange sight, but a local told me that this was a trace of a recent tree-avalanche in the Sound. This sounded even weirder, but the explanation was rather simple. The mountains around the Sound are densely wooded, but they are also extremely steep, so basically there is no soil. All the trees are clinging to each other, their roots forming a dense tangled mat. But every now and then the weight of the trees become too big, and part of these big forest-mats will slide down the side of the mountain as a tree avalanche, and sail of into the sunset.

Could it be that something similar sometimes take place in other parts of the world? Maybe not as avalanches, but on a smaller scale? Imagine sailing along, seeing this island with a few trees, going ashore, and suddenly realising that the whole outfit is sinking because you have put your big feet on it? And as we all know, trees can't just sail along on their own, so of course they must have been growing on the back of some monstrous animal!

Farfetched perhaps, but I do have six separate eyewitness accounts of sailing trees seen on Finnish lakes.

The lake troll

There are some other monsters in the heavyweight division in Finland. In Lake Opp, there is a monster which I originally thought was nothing but a folk tale, a lake-troll that will regularly go on a rampage and harass fishermen and destroy their nets. It is even said that the troll is bound to the lake by two rune-stones at opposite ends of the lake. It can only show itself when the stones are obscured by fog.

Fog is the perfect weather for imagining things. There is something about a fog that makes even the smallest and most everyday object look large and menacing, so any sighting of the troll in foggy weather could be otters, beavers, birds, bears, moose or just about anything out on, or in, the water.

But there are sightings on clear days as well, and they are kind of strange. Most of the sightings are from very hot and humid summer days, which is a good indication of something alive being behind all of it. But the animal is rather unusual, allegedly with a very big head, a bristly beard or moustache and a short thickset neck. To me that sounds almost like a form of walrus or perhaps a seal of some kind. The only thing is that lake-living seals are usually smaller than their marine counterparts, and walruses are exclusively marine. So what is it?

The terrible *tursus*

A monster built somewhere along the same lines, although on an even bigger scale, is the *tursus*. This creature is a legendary monster you can read about in various folk-legends. It has the torso of a man, but the backend of a whale or big seal, arms like a human - sometimes just one arm - the head of a giant walrus with large ears, and is sometimes dressed in sealskins. Legend? Most surely. Imagination? Perhaps, but a real creature may lurk somewhere behind all the trimmings.

The *tursus* does look like a giant garbled version of a walrus, which is a bit strange as there are definitely no walruses in the Baltic, and Finland hasn't got a coastline towards the Arctic Sea. However, that of course does not rule out that in earlier times traders, hunters or other strange wandering folk brought stories about these giant sea-mammals to the Finns, and perhaps sold them the odd tusk or two.

A few minor details do worry me though, for example the large ears, but I am inclined to put that down to artistic license. There has, in fact - in later years - been a few sightings of something that might have been the *tursu,* which of course sends my whole argument above down the drain, because this animal was in fact seen in the Baltic, off the city of Turku, in 1976 and 1978. Now as everybody who is worth his or her salt knows, 1976 was a very productive year cryptozoologically speaking. Strange animals were showing themselves all over the place, and apparently the area around Finland was no different than the rest of Europe.

So we cast a beady and somewhat bloodshot eye back towards 1976, when a Finn and a Swede were out on a boating trip off Turku. It was in the middle of July, and on a blistering hot day. The two friends had planned to do a spot of fishing, but when they got out there all they could muster enough energy for was to lie in the sun, talk and drink a beer or two. Suddenly all hell broke loose. A small group of gulls sitting on the water suddenly flew off, screaming their heads off, and something bumped into their boat so violently that things fell down, beer-bottles tumbled, and general confusion ensued.

The Finn, who prefers to remain anonymous, but who is a policeman in real life, almost fell out of his deckchair. And the Swede, who was lying on the deck, got a severe bump on the nose when a tray of sandwiches fell on his face.

The two men jumped to their feet thinking another boat had rammed them, but all they could see was a broad, wet, glistening dark greyish brown back of a large animal of some kind, just diving under the surface.

> "It was absolutely huge – we only saw a tiny part of its back, because it was diving, but that was something like 2 by 3 metres. My first thought was that a large whale had blundered all the way into the innermost part of the Baltic, but when it actually dived, it didn't show the broad flukes of a whale, but something that looked like a very large pair of flippers like you would see at the back end of a sea-lion, only much larger. I would say something like a metre across. I ran for my camera, but before I had a chance to take a picture, the animal had disappeared. I think it surfaced again about ten minutes later, but then it was several hundred metres off, and though it was a very calm day, all we could see was a big squarish head of some kind."

My best guess is a whale of some kind, perhaps a killer whale severely of its beaten track, but apart from that, I have really no good idea as to the identity of this disturber of the peace.

The merfolk

There are merfolk aplenty in Finnish waters, not just in the sea like in so many other countries, but also in lakes, stories of which tend to bring Harry Potter and the lake at Hogwarts to mind. Ugly looking creatures with strange screeching voices. None of the sightings I have been able to locate are new though, with one notable exception, but I'll save that one for last. They seem to be firmly rooted in the "Once-upon-a-time" realm - strange watery creatures living just below the surface, just beyond reach, but demanding a sacrifice every now and then. They have been seen for sure, mostly in olden days, sitting on rocks at the coast or poking a strange be-whiskered head up through the water-lilies on a quiet lake, but there are no definite sightings. There is nothing with: on this date, Mr. So-and-so saw this creature, so for now - or at least until I, or somebody else - locate some more precise sources, that's where the Finnish merfolk remain – shrouded in the mists of legend.

The man in the harbour

But just to underline the fact that the days of legends are far from over, we will leave Finland with this little story.

Two friends are standing on the deck of a cruise ship in Helsinki Harbour. It is rather late, close to midnight, but as it is the beginning of July 2007, it's not really dark. So, when a swimming man suddenly appears next to the ship, the two guys – both middle-aged business-men from Denmark on a vacation – can see him quite clearly.

"It wasn't just a normal person swimming up to the boat. He just appeared from below, like a submarine slowly rising, looked around very slowly and cautiously for a while, and then started – well, playing in the water is the best way I can describe it. It was like watching a trained dolphin. He turned and twisted and jumped clear out of the water several times. And after a few minutes disappeared as quickly and quietly as when he appeared."

"He looked to be the size of a normal man, but his skin had a strange yellowish colour and looked very rough, like he had scales or warts all over his body. He had a small triangular back fin, and very big flippered hands and feet. I have never, not even in the movies, seen anything like it, and I hope I never will again. It was creepy, and I had trouble sleeping for days afterwards."

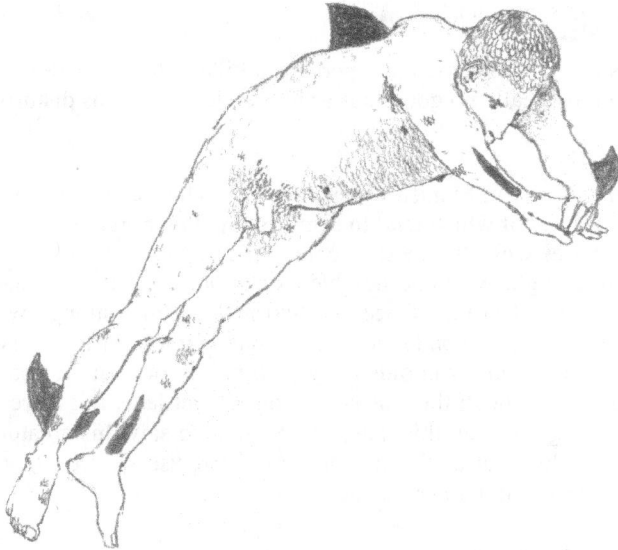

ESTONIA

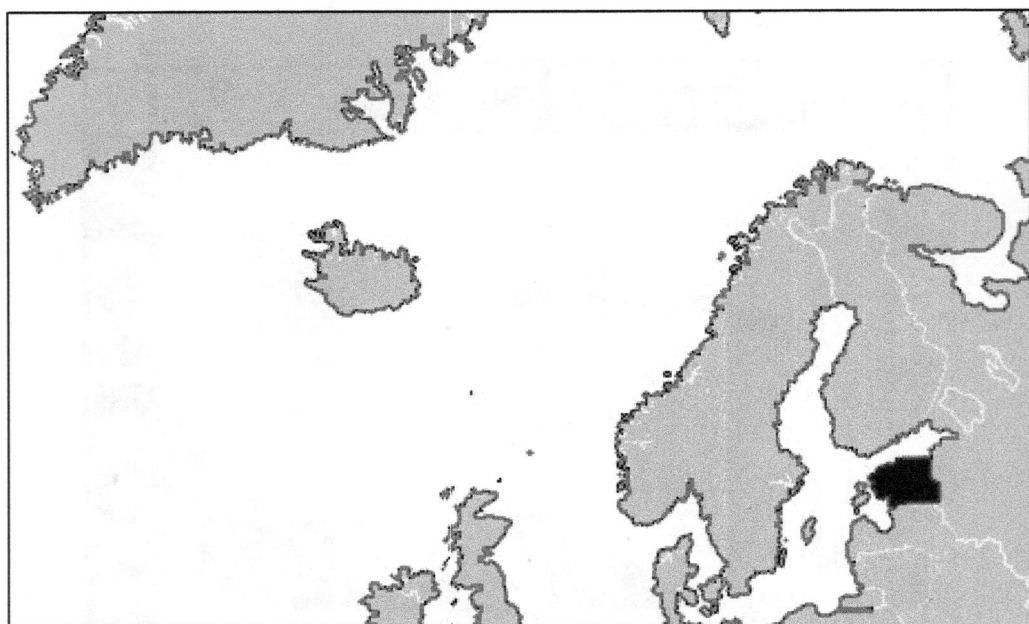

The Baltic countries are full of myth and legends, of heroes and brave knights of all descriptions, so the countries are steeped in dramatic tales of brave souls fighting terrible monsters. But amidst all the gory stuff, there are in fact stories which point to real life creatures being behind it all.

Pure bull (and cow)

Most of the countries in northern Europe have stories about merpeople or lake-people and their wayward cattle. There is always one or two of them doing a runner, and going up onto dry land and making a spectacle of themselves. Sometimes it is only bulls (like in Denmark) but in other countries, it is usually cows. And ugly brutes they are too those bulls and cows, but they are also prime examples of "looks aren't everything", as the cows are first class, perhaps even beyond first class, milkers, and are, therefore, the offspring of a merbull and an ordinary cow. Lucky is the farmer with cows like this.

But nothing lasts for ever, so eventually some day the merfolk will turn up and demand their animals back, and woe unto you if you haven't treated them properly.

For some reason, these stories are extremely common in Estonia. In this country these water cattle and their owners can even be found in lakes. And the descendants of the mating between mercattle and ordinary land cattle, allegedly live all over the country to this day, and you can sure see some ugly cows here and there.

But what kind of cryptozoological aspect is hidden in all this you may ask. Perhaps nothing, but I do have a bunch of sightings of cows, usually small, skinny and bony, suddenly appearing in some areas next to lakes, wandering about for a couple of days, and disappearing again without a trace – in one case leaving three-toed footprints behind.

Heaps of horses

Perhaps they were not really cows but waterhorses, something Estonia has aplenty. They leave three-toed footprints behind as well, and apparently the Estonian ones are of a particularly vicious and ill-tempered variety. They demand sacrifice on a regular basis, and if not properly serviced in this department will help themselves to whatever number of children they deem suitable. In some areas the regular sacrifices to waterhorses went on well into at least the early part of the 19th Century and perhaps even longer.

I cannot help but wonder why stories of waterhorses are so common in so many countries. Are they perhaps some kind of bogeymen-stories – don't go near the water, or the waterhorse will take you. This could go a long way in explaining why the waterhorses prefer children as victims. They are much more prone to go wandering off into dangerous situations. And perhaps in a land like Estonia with its multitude of lakes and swamps, the waterhorses are especially common for this particular reason.

Enter the hairy fishes
In a few lakes at various places in Estonia, for some reason especially around the city of Viitua, it is sometime possible to meet up with some rather strange looking, and extensively hairy yellow fish. They only appear in some years, and then only in certain seasons. The Icelandic hairy *lodsilungur* immediately springs to mind. I would imagine that something similar is happening here in Estonia. In some years a set of special circumstances ensures that various fish living in the lakes are attacked by some form of fungus that forms hair-like mats on the bodies of the fish. It would be interesting to catch one of these - apparently some were seen in 2002 - and subject them to closer scrutiny - perhaps a form of cryptofungus is behind all of this.

The Creature from the Black Lake
No, I am not talking about a 50s horror-movie. I am talking about a lake, which is strangely enough also in the vicinity of Viitua like the ones above, where nothing is ever caught. Nothing but rotten logs and foul-smelling ooze.

At one time this lake was a place where the fishing was good, but some time in the late 1700s or perhaps early 1800s – the actual time is not certain in any way – the local fishermen caught a very strange beast indeed. It was described as half fish, half hairy "thing", and smelled something awful. And to add insult to injury, the weather immediately took a turn for the worse when the net containing this thing landed on deck.

As the fishermen debated what to do with this hellish creature, the swell rose steadily higher, until they were all in fear of their lives. Finally, they decided to throw the creature over the side. This was duly done, and immediately the water went calm again. The sun came out, the birds started singing, and it was as if the creature had never been. Good news for the fishermen one would imagine, but alas no. They lived to go ashore and tell their story, but since then it has been impossible to catch anything in the Black Lake.

Now I think that what we have here is a case of two separate happenings being mixed together to form a legend, or one being used as explanation for the other.

It is quite easy to explain the lifelessness of the Black Lake as the effect of some kind of pollution, perhaps dating back to the turn of the last centuries or to something that happened during the Soviet age. It is a bit more difficult to explain the actual catch, but a dead and partly decomposed animal of some kind? A very large and perhaps scary looking wells catfish? Add a few embellishments – rising waves and black clouds, and cue the scary music.

And that, as they say is probably that, unless of course the fishermen somehow had invoked "The Ancient One", Estonia's biggest, baddest, and most terrifying monster.

The Ancient One

The description of The Ancient One is enough to scare anyone out of their wits, as it is a monstrous fish-shaped creature, but one that walks on land on thick powerful legs. It has a maw the size of the gates of Hell, and a massive saw-toothed ridge on its back.

Now here *is* a monster, if ever there was one. The stories of The Ancient One go back a long, long way, but I do have a single modern sighting from 1987, when two people in a car saw what they described as a dinosaur walking across the road near Viitua and disappearing into a lake.

Now what, pray, are we to make of this? Central African creatures like mokele-mbembe do spring to mind. Or maybe old and garbled sightings about meetings with crocodiles? Perhaps even in far and foreign lands. Stories that somehow were transported to Estonia just as stories of black dogs have moved from Europe to the United States, and the *chupacabras* has taken the trip from Puerto Rico to America. If anyone out there has ever set eyes on The Ancient One give me a call.... Please!

The Ancient One

LATVIA

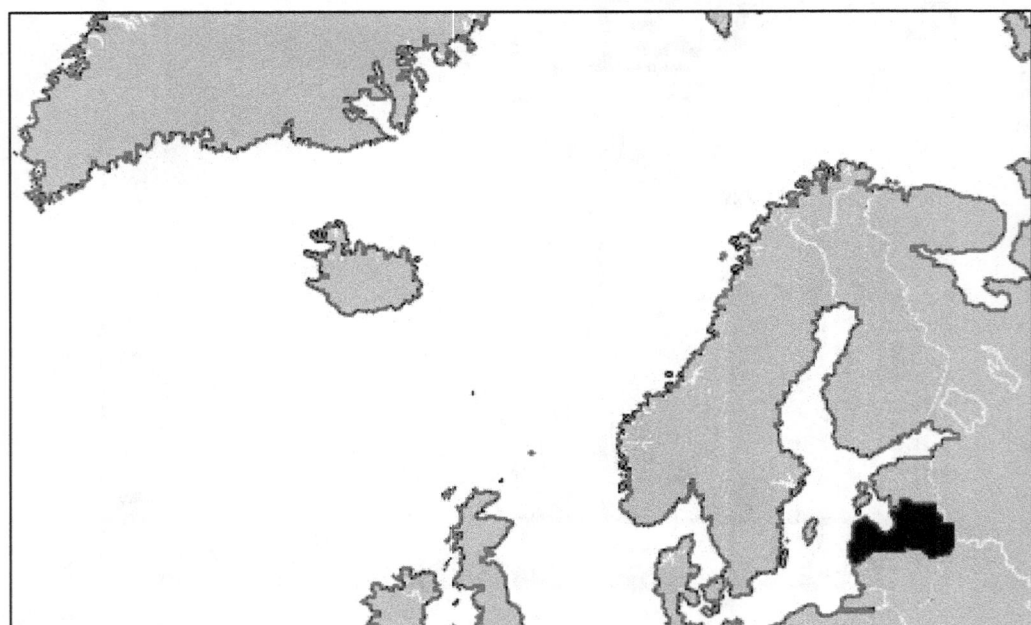

Latvia is not a country where information on monsters and similar stuff is easy to come by. If you just walk around like any bloody tourist and start posing strange questions to people that you meet on the street, they tend to look at you with a fair amount of scepticism and mistrust. This is something for which I do not blame them, but thanks to a friend - who for some years assisted the Latvian Government in connection with various conservation projects - a few stories have been coming my way. It is not much, I will be the first to admit, but even the smallest amount from a country so cryptozoologically neglected should be interesting, and since the number of tourists to the Baltic countries is steadily growing, and a lot of them come for the wildlife, I have high hopes for further stories emerging from Latvia.

Here be ... nothing?

Latvia has a fairly short coastline along the Baltic Sea, and not much of a tradition for seafaring, so perhaps it is not that strange that I have very few stories about sea-monsters from Latvian waters. Most of these are very vague along the lines of "something large swimming at or just below the surface", but with no actual details as to the size, shape and colour of this 'thing'. The only remotely interesting story concerns a stranding, sometime in the 1700s, of a very large and hairy beast, with four short legs, a long tapering tail and a small head on a long neck.

This to me sounds very much like the remains of a decomposing big shark, probably a basking shark. Apparently this creature reeked something awful, with a smell that burned your nose and stung your eyes. Having dissected several dead sharks, although Greenland sleeper sharks, I can testify to the fact that if a shark is ripe, the smell is rather overpowering.

Of heroes and monsters

If we move inland to some of the country's many lakes or swampy areas, there are richer pickings and more excitement to be had. But before we venture any further, we have to get the Latvian national hero Lacplesis - The Bearslayer – out of the way. He was a close colleague of Gilgamesh and numerous other monster-slaying heroes all over the world, and apparently he once roamed the countryside killing dangerous beasties right, left and centre. And bears of course, (hence the nickname), which he of course killed with his bare hands. But he also tangled with giants and a few lake-monsters here and there. The most famous of which was a gigantic three-headed thing living in the Daugara River, which was disposed of quickly and efficiently. This story has a certain life of its own, and you can hear it told in several parts of the country. So all stories featuring monsters being three-headed, or similarly endowed in the cephalic department, I have regarded as variations on this heroic folktale. But if nothing else, the stories have shown us that, like the old medieval saying still goes, here be monsters.

If you go down to the lake today...

You might not see a thing, but then again sometimes you do. Latvia is quite well endowed when it comes to lakes and wetlands of all kinds. There is plenty of opportunity to get your feet wet, and to be eaten alive by billions of voracious mosquitoes should you be foolish enough to venture out there in the middle of summer. However, luckily for me some people have braved the elements, and the dipteran terror and gone out, and they have seen things – big things, small things, and not to mention strange things. Some may be due to sunstroke (it does get hot out there) or mirages, or perhaps are induced by the slight hysteria which I at least always feel rising when I am cast in the role of someone's lunch. But some things have most certainly been alive out there.

The snakes are here

The eastern part of Latvia is especially strong on the lake front. There are a lot of small, medium-sized and large lakes here. Most of them are fairly shallow, but there are some substantial bodies of water here and there.

I have no real idea as to how many lake-monsters are splashing about in the country, but my feeling is that there are quite a lot, although the number of actual sightings I have been able to collect is fairly small. But I do have some fifteen sightings from four different lakes: Kurganova, Lubans, Cirmas and Raznaz, and if they are any indication of what is actually out there, Latvia is a cryptozoological treasure trove.

I will not go into detail with the actual sightings, as all of them are of the well-known "giant-snake" type creature, which I have looked at several times in this book already. We might be

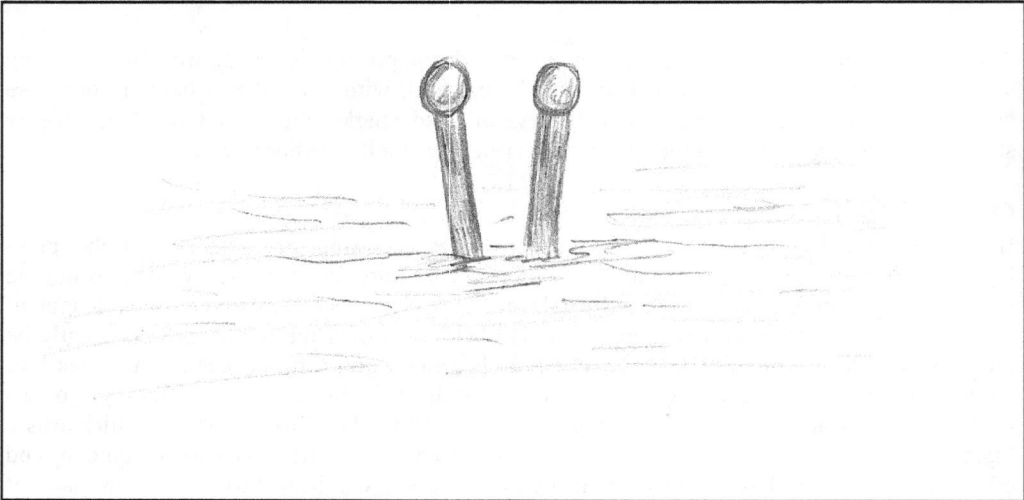

These strange antenna-like objects have been seen in several lakes in Latvia - the antenna of the fireslug perhaps? (Eyewitness drawing)

dealing with the same giant sterile eels or something similar here, but then again, we might not. At least one of the animals had a long saw-toothed ridge along the back, which sounds distinctly more reptilian to me. Perhaps some of the stories you hear in ancient legends about dragons sleeping in or underneath lakes are not so far from the truth after all.

The Big One

The monster in Lake Usmas in the western part of the country is a bit different. This is in an area filled with swamps and many smaller lakes. The creature seems to be an entirely different species of animal. I have corresponded with a man who lived in the area when he was a child, and he claimed his parents always warned him to be wary of a creature they called "The Big One". He never saw anything, but told me about a friend who claimed to have seen a very large turtle swimming in the lake one day in the middle of summer. There was nothing really special about the creature, apart from its rather humongous stature. It was at least the size of a big sea-turtle, with a big square head, and a fearsome looking hooked beak.

This sounds very much like an alligator snapping turtle, although that species is of course an American one. But could there be something similar living in Eastern Europe and Asia? There is always the possibility of an escapee, but according to my source, the stories about "The Big One" goes back to at least the late 1800s, since his parents had heard the same warning when they were small from their parents. But if there is some kind of giant snapping turtle living in these areas, I suggest that the warning should be taken very seriously as those are not animals to be trifled with if you value your current number of fingers and toes.

The waterponies

Like so many other countries, Latvia has its fair share of stories about waterhorses and similar beings. The only difference when it comes to the Latvian ones, is the fact that they tend to be rather short and stocky, and a bit on the hairy side of things kelpie. Waterponies perhaps? But apart from their looks, they tend to behave in a similar vein to other waterhorses, that is every now and then luring a poor soul to his or her watery death beneath the waves, but in general just keeping to themselves if they are left alone.

The fire-slug

By far the strangest creature to emerge from the Latvian swamps is an animal with the suitably strange name of fire-slug, a creature that seems to be a distant relative of the Mongolian death-worm.

I heard the story of the fire-slug from a Latvian student who came to Denmark some years ago on a scholarship to study zoology. I met him one day at the university, and after having told him about my interest in strange creatures, he told about this weird creature from the swampy areas in the western part of the country.

The fire-slug is an extremely rare creature. It only shows itself very rarely and usually you only see its tracks, roughly 50-60 cm wide paths that go through the forest, looking like some-one has gone through with a flame thrower or a heavy duty weed-burner. And the charred and wilted remains of the plants on the tracks are covered with an evil smelling slime that will burn the skin if you touch it.

The animal making these tracks looks like a giant slug, some 3-4 metres in length, with two sets of antenna. The damage is supposedly done by the slime, which is very acidic and has a strong corrosive effect. The creature does not actually breathe fire or anything and is, as far as is known, fairly placid and non-aggressive, but should you touch a fire-slug, agony would most surely follow as it is covered in a layer of its own dangerous slime. In dangerous situations, the creature is even capable of squirting this slime about in a wanton and careless manner, and then I suggest you do not want to be around. Get some in your eyes, and you will probably end up blind.

What to make of that? Folklore, urban legend, or the student pulling my leg? (Or perhaps a

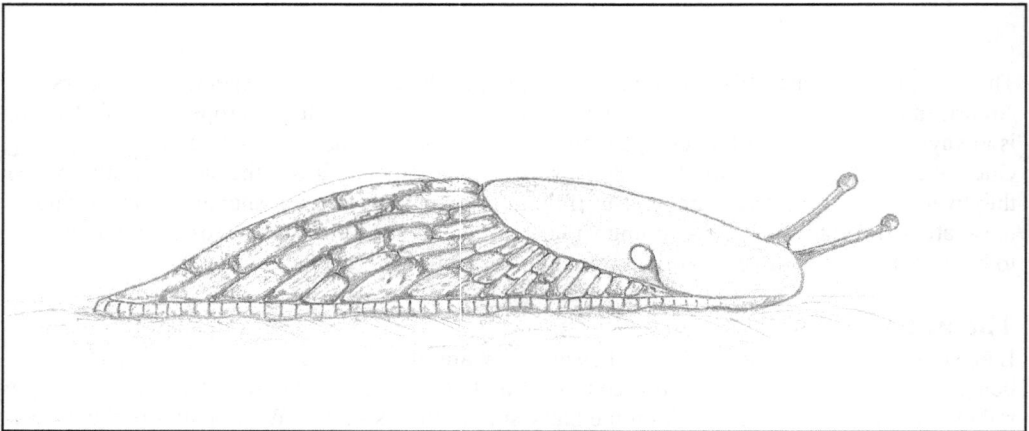

little of each?) I am sure that the last is not an option, since I have now heard the story from a couple of other sources, but as to the rest of it? I don't know. Some aspects of the story smacks of urban legend, and I am not even certain a slug that big would be biologically possible – but then again, why not? Using slime as a defensive mechanism sounds perfectly fine to me, and a poisonous and corrosive slime would most assuredly have a detrimental effect on any kind of plant-life. But I am still to find an eyewitness who has actually seen the slug or its tracks.

LITHUANIA

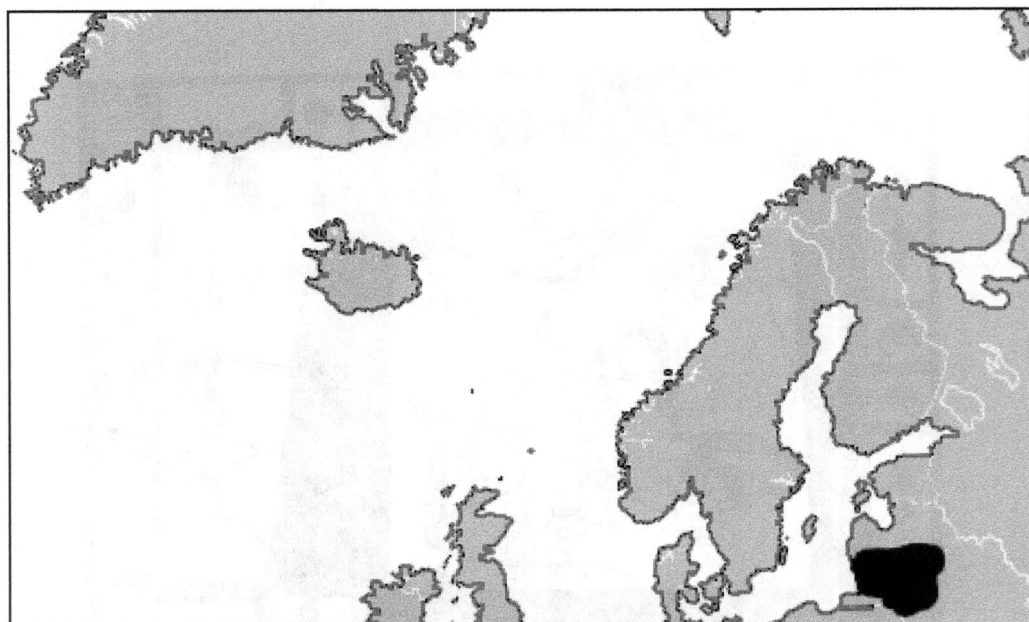

Lithuania is a country of many, many lakes, but not much to speak of when it comes to sea-monsters. The country's coastline is short, and the Baltic is not good for sea-monsters anyway. I have in fact not been able to find a single history, or a single observation of a sea-monster made by a Lithuanian or in Lithuanian waters. But when it comes to lakes and swamps, the game's very much afoot – to quote a famous detective.

The number of sightings is not great – or at least I have only been able to find a rather limited number, but the diversity is rather surprising.

The wandering worm
For starters, we are only going to get our feet slightly wet because we are going hunting for some very large and very strange worms seen in the forest south-west of Lithuania's capital city of Vilnius. I have only been able to find three sightings: one from "sometime in the 1920s", one from "a few years after the end of the Second World War", and one from 1976 (there is that year again). The sightings are remarkably similar. I have only been in personal contact with the eyewitness to the 1976 sighting.

The 1920s eyewitness was a duck-hunter, the next witness was a man looking for mushrooms, and in the third case, it was an art student looking for suitable scenery to paint. But I will let Jannis tell his story:

> "I had been out in the forest for about an hour, and was taking a break on a stump next to a swampy area between a pathway and a little lake. I was thinking about what I wanted to paint, when I suddenly heard a very strange "sticky" sound. It reminded me of the sound it makes, when my mother makes jam, and is pouring it from the pan into jars.
>
> "At first I couldn't see anything, but suddenly I realized something was moving among the trees in the swamp. It took a little while for my eyes to get used to the darkness among the trees, but then I realized I was looking on an enormous greyish white worm slowly creeping along on the ground, and making the strange noise. Unfortunately I couldn't get any closer – I was afraid of being caught in the mud, but I think the worm was something like 5 or 6 meters long, and perhaps 20 or 30 cm in width, and very flat, like a thick piece of paper. I had no camera, but I made a drawing of what I saw."

The drawing is not especially detailed, it is more like an impressionistic painting, but the white ribbon-like structure of the creature is quite clear.

The art student's impression of the strange flat white "worm".

My first thoughts when I heard this story was something like a giant flatworm or an enormous leech, but suddenly I realised that we have stories in Denmark of something rather similar.

Since medieval times people have talked about and feared the terrible "armyworm", an omen of war, pestilence and famine; a strange grey or white snakelike thing that suddenly appears in a forest somewhere, is seen by a few people, and usually disappears just as suddenly again.

Scary – huh?

Not really. The army worm has an interesting, but quite straightforward explanation. Enter the dark-winged fungus gnats, small members of the family Sciaridae. These small and completely harmless mosquito-like insects sometimes band together – or rather their larvae do – and crawl about in their thousands in a long snake-like column. The column rarely lasts for long, sometimes only a couple of hours, and then the larvae go their separate ways. I am quite certain, that this is what Jannis and the others have seen. Something quite rare and very interesting – but not a monster!

Tatzelwurm or what?

Next we move to the eastern part of the country, to the area around Trakan where there are a lot of small lakes. In this area, I have a number of sightings of an animal I can only describe as salamander-like or perhaps even tatzelwurm-like.

It has been seen on land as well as in the water, swimming rather slowly and lazily about or crawling along the bottom. It looks like a very large salamander some 40-50 centimetres in length, with a rather flat and powerful body and rather short and stumpy legs, quite unlike the often spindly legs of ordinary salamanders. It also lacks the prominent gills usually to be found on water-dwelling salamanders or newts. The colour is usually a uniformly dull greenish brown, sometimes with a few yellow or orange blobs here and there.

This all sounds extremely salamander-like to me. But when the critter walks on dry land, it does exhibit some distinctly unsalamander-like traits. It can apparently jump, or at least lift the front parts of the body up off the ground, and it is even said to emit a strange barking sound. The only animal I can think of to fit this description is some kind of salamander, although so unusual it must warrant at least a new species, should it ever be caught. But it has also made me wonder whether this creature is the actual role model for the medieval and more magic fireproof salamander? It is just a thought, but I have always found it hard to believe that a tiny fire-salamander should inspire such unusual legends.

An otter story (and no, it's not a typo)

The next story comes from Lake Metelys in southern Lithuania where, in the spring of 2002, a friend of mine was trying to take a picture of a ferruginous duck. He had been busy fiddling with his camera and snapping away when the head of an otter suddenly popped up among the various ducks on the lake. Nothing strange there, except for the fact that my friend swears the animal had a head the size of a large pumpkin. After all, there was plenty of ducks around for comparison. To quote:

> "It looked like it was capable of swallowing one of the ducks in a single gulp."

Unfortunately, the animal disappeared as quickly at it had showed itself in the first place. All together it was only at the surface for a couple of seconds, and my friend only had time to grab his binoculars for a quick look. Alas, the critter never reappeared, but I trust my friend explicitly, and what's more, when not gallivanting around the globe watching birds and butterflies, he is a mammalogist by profession. Granted he works with rodents, but he does know an otter when he sees one, even a monstrous giant. He tried to ask around locally, but could not get any more information pertaining to this monster. I have tried numerous times to get more information, but with no greater luck.

And the animal? I haven't got the faintest idea to tell you the truth. An ordinary otter, whose head got distorted by some kind of mirage phenomenon is all I can think off, but my friend is adamant there was no distortion in the air that day.

Eech! A leech!

Lake Dusia is the largest lake in southern Lithuania with a surface area of some 23 km^2, but with an average depth of only 14.6 m. I have heard a few vague stories about somebody who had, at some unspecified time, seen something snake-like swimming about on the surface, but unfortunately there are no more details. A bit more specific are the sightings of two extremely large leeches in a flooded area close to the lake in the spring of 1999. This story came to me courtesy of two Danish birdwatchers who saw the leeches swimming past. There was nothing particularly special about their sighting, except for the fact that the leeches were some 1.5 metres in length. Not the kind of thing you would like to fasten itself to your lower calf.

Now, I know there are some rather longish leeches in tropical areas, but 1.5 metres is ridiculous, and that was when swimming. When fully stretched they would probably be something like 2.5 metres. The things were very dark green, with a few yellow spots on what must have been the back.

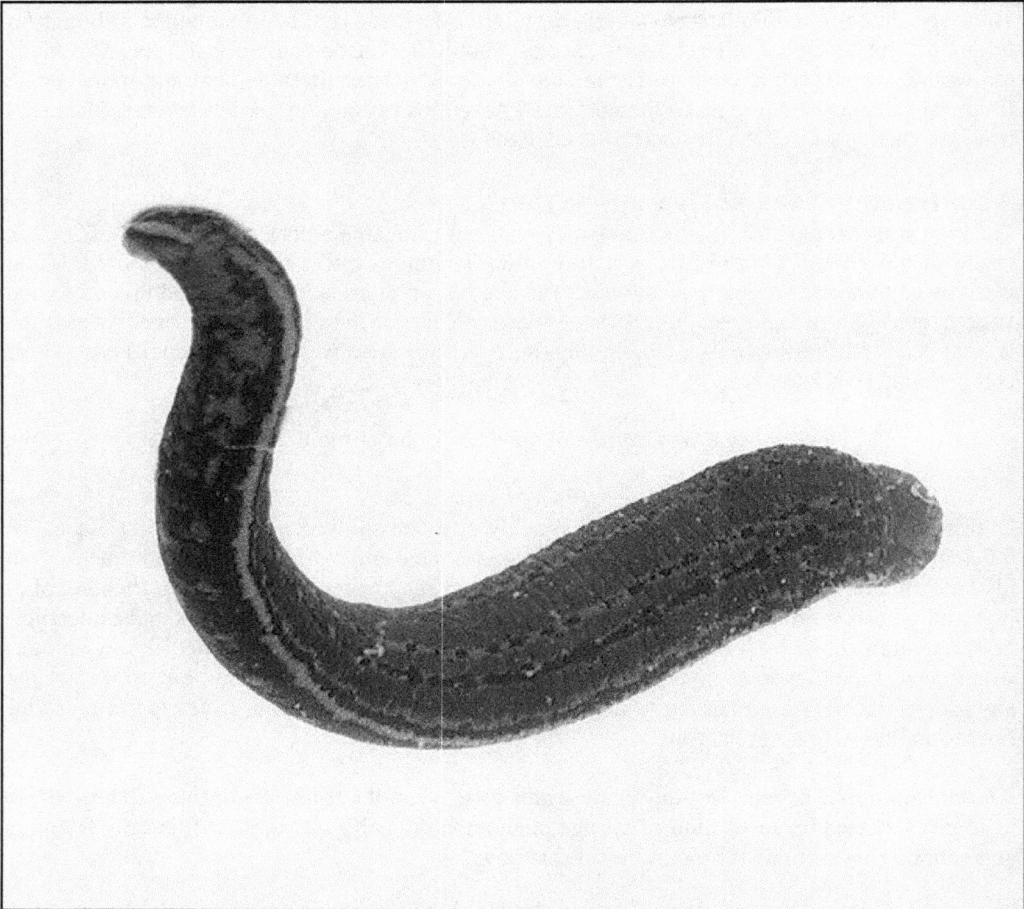

Were they a couple of very, very old leeches that never got around to breeding and just kept growing and growing? Or a completely new species? I wish I knew!

Up periscope – or something

Lake Pidziulis can be found just outside the little city of Daugai in southern Lithuania. On the surface (ha ha!) there is nothing special about the lake, but there may be something rather strange lurking in its depths.

In my files I have four sightings, all from the 1990s, of something described as two long poles (about 60-80 centimetres in length) with "blobs" the size of grapefruit on the tips. They suddenly rose from the waters of the lake, usually some 20-30 metres from the shore, and started waving slowly back and forth like the antenna of a very large slug. After a couple of minutes, the "poles" would slowly sink again and disappear like a couple of twin periscopes on a mini-sub.

As for the identity of that thing, it is another one of those where I have absolutely no idea – but perhaps – just perhaps, this has something to do with the Latvian fire-slug (please refer to the Latvian chapter for this).

IS THERE A MONSTER IN THE AUDIENCE, OR PERHAPS AN EYEWITNESS?

Northern Europe is an extremely diverse area in more ways than one, even cryptozoologically, as these pages hopefully have shown. I hope they have been fun to read, and I very much hope that they have piqued your curiosity. Now get out there, in the swamps and bogs, and along the lakes and the rivers. Keep your eyes open, and if you see anything, or hear a good story please let me know!

Sources and further reading

Ashton, John: *Curious Creatures in Zoology,* Cassell, London 1890
Barber & Riches: *A dictionary of fabulous beasts.* Boydell press 1971.
Baring-Gould, S.: *Curious Myths of the Middle Ages*, John B.Alden Publishers,
        New York 1885
Briggs, Katherine: *An encyclopedia of fairies.* Pantheon Books 1976
Briggs, Katherine: *The vanishing people.* Batsford 1978.
Eberhart, G.M.: *Mysterious Creatures*, ABC Clio 2002
Egede, Hans: *Det gamle Grønlands nye perlustration eller Naturel-Historie,*
        Copenhagen 1741 (in Danish)
Fortean Times 43: 25
Fortean Times 46: 29
Holme, F.E.: *Natural History Lore and Legend*, Bernard Quaritch, London 1895
ISC Newsletter 4/3: 10
Magnus, Olaus: *Historia de Gentibus Septentrionalibus*, Rome 1555 (in Latin)
Meurger & Gagnon: *Lake Monster Traditions.* Fortean Tomes 1988.
Newton, Michael: *Encyclopedia of Cryptozoology*, McFarland & Company 2005
Nordisk Familjebok, Stockholm 1904-1926 (in Swedish)
Notes & Queries Dec. 17th. 1859
Oudemans, A.C.: *The Great Sea-Serpent*, Brill, Leiden 1892
Paxton, C.G.M. & R.Holland: Was Steenstrup Right? A New Interpretation of the
        16th Century Sea Monk of The Øresund, *Steenstrupia* 29(1), 2005: 39-47
Paxton, C.G.M., Knatterud, E. and S.L. Hedley: Cetaceans, sex and sea serpents: an analysis
        of the Egede accounts of a "most dreadful monster" seen off the coast of
        Greenland in 1734. *Archives of Natural History* 32(1): 1-9, 2005
Pontoppidan, Erik: *Norges Naturlige Historie*, Copenhagen 1753 (in Danish)
Rasmussen, Knud: *Myter og sagn fra Grønland*, Sesam, Copenhagen 1994 (in Danish)
Rose, Carol: *Giants, monsters and dragons. An encyclopedia of folklore, legend and myth.*
        ABC-CLIO 2000.
Samuelsen, Espen: In search of Norway's Nessie, "Rømmie". *Fortean Times* January 2002.
Skjelsvik, Elizabeth: Norwegian Lake and Sea Monsters, *Norveg 7*, 1960: 29-48
Steenstrup, J. J. S: *Om deni Kong Christians IIIs Tid I Øresundet fangne Havmand*, Sømunken
        kaldet, Almenfattelige Naturskildringer 1854.
Thomas, Lars: *Det mystiske Danmark*, Aschehoug, Copenhagen 2007 (in Danish)
Thomas, Lars: *Det mystiske Danmark – 2. samling*, Aschehoug, Copenhagen 2008 (in Danish)
Thomas, Lars: *Det uforklarlige*, Sesam, Copenhagen 2002 (in Danish)

Thomas, Lars: *Flere uforklarlige fænomener*, Sesam, Copenhagen 2003 (in Danish)
Thoresen, Magdalene: *Billeder fra vestkysten av Norge*, Gyldendal 1872
Varner, Gary R.: *Mermaids & Water Spirits*, www.authorsden.com
Water monsters: Greenland, *Fortean Times* 46: 29

hhtp://hem.passagen.se/gryttie/about.html
www.americanmonsters.com
www.lakedragons.livingdinos.com
www.mjoesormen.no
www.storsjoodjuret.com

The Newspaper Department in the Royal Library in Copenhagen

INDEX

Places:

People:

Ordinary animals mentioned in the text with bearing on the monsters:

Monsters:

STILL ON THE TRACK OF UNKNOWN ANIMALS

The Centre for Fortean Zoology, or CFZ, is a non profit-making organisation founded in 1992 with the aim of being a clearing house for information, and coordinating research into mystery animals around the world.

We also study out of place animals, rare and aberrant animal behaviour, and Zooform Phenomena; little-understood "things" that appear to be animals, but which are in fact nothing of the sort, and not even alive (at least in the way we understand the term).

Not only are we the biggest organisation of our type in the world, but - or so we like to think - we are the best. We are certainly the only truly global cryptozoological research organisation, and we carry out our investigations using a strictly scientific set of guidelines. We are expanding all the time and looking to recruit new members to help us in our research into mysterious animals and strange creatures across the globe.

Why should you join us? Because, if you are genuinely interested in trying to solve the last great mysteries of Mother Nature, there is nobody better than us with whom to do it.

Members get a four-issue subscription to our journal *Animals & Men*. Each issue contains nearly 100 pages packed with news, articles, letters, research papers, field reports, and even a gossip column! The magazine is Royal Octavo in format with a full colour cover. You also have access to one of the world's largest collections of resource material dealing with cryptozoology and allied disciplines, and people from the CFZ membership regularly take part in fieldwork and expeditions around the world.

The CFZ is managed by a three-man board of trustees, with a non-profit making trust registered with HM Government Stamp Office. The board of trustees is supported by a Permanent Directorate of full and part-time staff, and advised by a Consultancy Board of specialists - many of whom are world-renowned experts in their particular field. We have regional representatives across the UK, the USA, and many other parts of the world, and are affiliated with other organisations whose aims and protocols mirror our own.

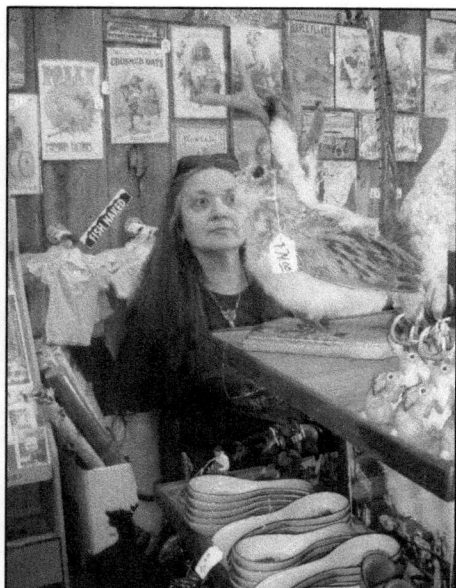

You'll find that the people at the CFZ are friendly and approachable. We have a thriving forum on the website which is the hub of an ever-growing electronic community. You will soon find your feet. Many members of the CFZ Permanent Directorate started off as ordinary members, and now work full-time chasing monsters around the world.

Write to us, e-mail us, or telephone us. The list of future projects on the website is not exhaustive. If you have a good idea for an investigation, please tell us. We may well be able to help.

We are always looking for volunteers to join us. If you see a project that interests you, do not hesitate to get in touch with us. Under certain circumstances we can help provide funding for your trip. If you look on the future projects section of the website, you can see some of the projects that we have pencilled in for the next few years.

In 2003 and 2004 we sent three-man expeditions to Sumatra looking for Orang-Pendek - a semi-legendary bipedal ape. The same three went to Mongolia in 2005. All three members started off merely subscribers to the CFZ magazine. Next time it could be you!

We have no magic sources of income. All our funds come from donations, membership fees, and sales of our publications and merchandise. We are always looking for corporate sponsorship, and other sources of revenue. If you have any ideas for fund-raising please let us know.

However, unlike other cryptozoological organisations in the past, we do not live in an intellectual ivory tower. We are not afraid to get our hands dirty, and furthermore we are not one of those organisations where the membership have to raise money so that a privileged few can go on expensive foreign trips. Our research teams, both in the UK and abroad, consist of a mixture of experienced and inexperienced personnel. We are truly a community, and work on the premise that the benefits of CFZ membership are open to all.

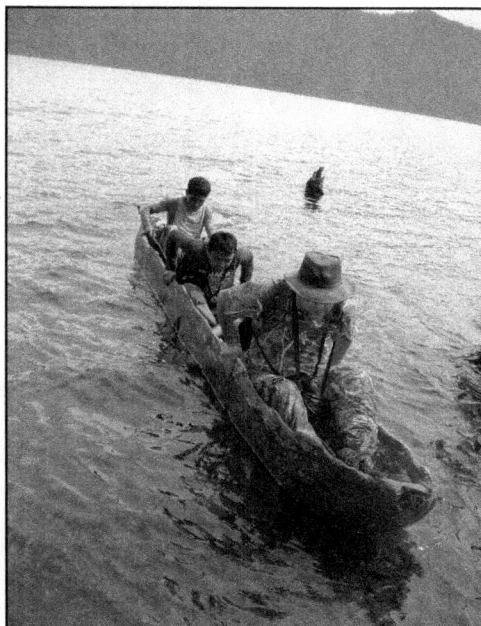

Reports of our investigations are published on our website as soon as they are available. Preliminary reports are posted within days of the project finishing.

Each year we publish a 200 page yearbook containing research papers and expedition reports too long to be printed in the journal. We freely circulate our information to anybody who asks for it.

We have a thriving YouTube channel, CFZtv, which has well over two hundred self-made documentaries, lecture appearances, and episodes of our monthly webTV show. We have a daily online magazine, which has over a million hits each year.

Each year since 2000 we have held our annual convention - the Weird Weekend. It is three days of lectures, workshops, and excursions. But most importantly it is a chance for members of the CFZ to meet each other, and to talk with the members of the permanent directorate in a relaxed and informal setting and preferably with a pint of beer in one hand. Since 2006 - the Weird Weekend has been bigger and better and held on the third weekend in August in the idyllic rural location of Woolsery in North Devon.

Since relocating to North Devon in 2005 we have become ever more closely involved with other community organisations, and we hope that this trend will continue. We have also worked closely with Police Forces across the UK as consultants for animal mutilation cases, and we intend to forge closer links with the coastguard and other community services. We want to work closely with those who regularly travel into the Bristol Channel, so that if the recent trend of exotic animal visitors to our coastal waters continues, we can be out there as soon as possible.

Apart from having been the only Fortean Zoological organisation in the world to have consistently published material on all aspects of the subject for over a decade, we have achieved the following concrete results:

• Disproved the myth relating to the headless so-called sea-serpent carcass of Durgan beach in Cornwall 1975
• Disproved the story of the 1988 puma skull of

Lustleigh Cleave
- Carried out the only in-depth research ever into the mythos of the Cornish Owlman.
- Made the first records of a tropical species of lamprey
- Made the first records of a luminous cave gnat larva in Thailand
- Discovered a possible new species of British mammal - the beech marten
- In 1994-6 carried out the first archival fortean zoological survey of Hong Kong
- In the year 2000, CFZ theories were confirmed when a new species of lizard was added to the British List
- Identified the monster of Martin Mere in Lancashire as a giant wels catfish
- Expanded the known range of Armitage's skink in the Gambia by 80%
- Obtained photographic evidence of the remains of Europe's largest known pike
- Carried out the first ever in-depth study of the ninki-nanka
- Carried out the first attempt to breed Puerto Rican cave snails in captivity
- Were the first European explorers to visit the `lost valley` in Sumatra
- Published the first ever evidence for a new tribe of pygmies in Guyana
- Published the first evidence for a new species of caiman in Guyana
- Filmed unknown creatures

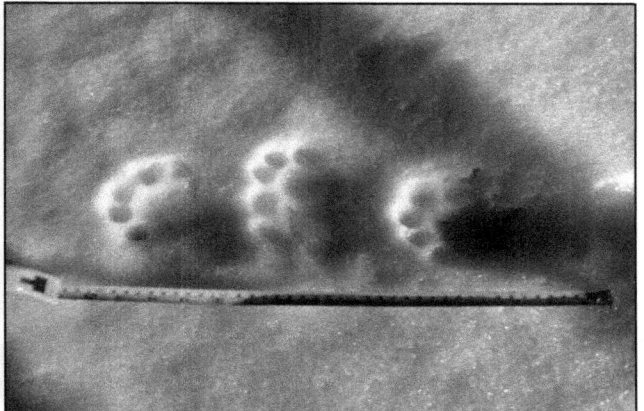

on a monster-haunted lake in Ireland for the first time

• Had a sighting of orang pendek in Sumatra in 2009

• Found leopard hair, subsequently identified by DNA analysis, from rural North Devon in 2010

• Brought back hairs which appear to be from an unknown primate in Sumatra

• Published some of the best evidence ever for the almasty in southern Russia

CFZ Expeditions and Investigations include:

• 1998 Puerto Rico, Florida, Mexico (Chupacabras)
• 1999 Nevada (Bigfoot)
• 2000 Thailand (Naga)
• 2002 Martin Mere (Giant catfish)
• 2002 Cleveland (Wallaby mutilation)

- 2003 Bolam Lake (BHM Reports)
- 2003 Sumatra (Orang Pendek)
- 2003 Texas (Bigfoot; giant snapping turtles)
- 2004 Sumatra (Orang Pendek; cigau, a sabre-toothed cat)
- 2004 Illinois (Black panthers; cicada swarm)
- 2004 Texas (Mystery blue dog)
- Loch Morar (Monster)
- 2004 Puerto Rico (Chupacabras; carnivorous cave snails)
- 2005 Belize (Affiliate expedition for hairy dwarfs)
- 2005 Loch Ness (Monster)
- 2005 Mongolia (Allghoi Khorkhoi aka Mongolian death worm)

- 2006 Gambia (Gambo - Gambian sea monster , Ninki Nanka and Armitage's skink
- 2006 Llangorse Lake (Giant pike, giant eels)
- 2006 Windermere (Giant eels)
- 2007 Coniston Water (Giant eels)
- 2007 Guyana (Giant anaconda, didi, water tiger)
- 2008 Russia (Almasty)
- 2009 Sumatra (Orang pendek)
- 2009 Republic of Ireland (Lake Monster)
- 2010 Texas (Blue Dogs)
- 2010 India (Mande Burung)

For details of current membership fees, current expeditions and investigations, and voluntary posts within the CFZ that need your help, please do not hesitate to contact us.

The Centre for Fortean Zoology,
Myrtle Cottage,
Woolfardisworthy,
Bideford, North Devon
EX39 5QR

Telephone 01237 431413
Fax+44 (0)7006-074-925
eMail info@cfz.org.uk

Websites:

www.cfz.org.uk
www.weirdweekend.org

THE WORLD'S

WEIRDEST

PUBLISHING

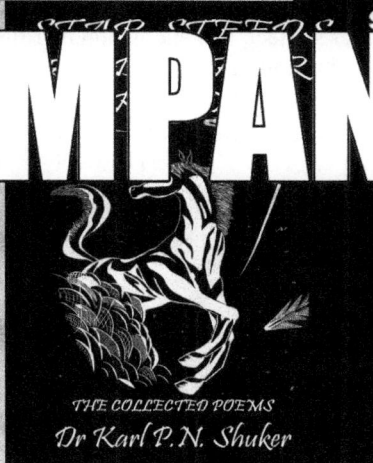

COMPANY

HOW TO START A PUBLISHING EMPIRE

Unlike most mainstream publishers, we have a non-commercial remit, and our mission statement claims that "we publish books because they deserve to be published, not because we think that we can make money out of them". Our motto is the Latin Tag *Pro bona causa facimus* (we do it for good reason), a slogan taken from a children's book *The Case of the Silver Egg* by the late Desmond Skirrow.

WIKIPEDIA: "The first book published was in 1988. *Take this Brother may it Serve you Well* was a guide to Beatles bootlegs by Jonathan Downes. It sold quite well, but was hampered by very poor production values, being photocopied, and held together by a plastic clip binder. In 1988 A5 clip binders were hard to get hold of, so the publishers took A4 binders and cut them in half with a hacksaw. It now reaches surprisingly high prices second hand.

The production quality improved slightly over the years, and after 1999 all the books produced were ringbound with laminated colour covers. In 2004, however, they signed an agreement with Lightning Source, and all books are now produced perfect bound, with full colour covers."

Until 2010 all our books, the majority of which are/were on the subject of mystery animals and allied disciplines, were published by `CFZ Press`, the publishing arm of the Centre for Fortean Zoology (CFZ), and we urged our readers and followers to draw a discreet veil over the books that we published that were completely off topic to the CFZ.

However, in 2010 we decided that enough was enough and launched a second imprint, `Fortean Words` which aims to cover a wide range of non animal-related esoteric subjects. Other imprints will be launched as and when we feel like it, however the basic ethos of the company remains the same: Our job is to publish books and magazines that we feel are worth publishing, whether or not they are going to sell. Money is, after all - as my dear old Mama once told me - a rather vulgar subject, and she would be rolling in her grave if she thought that her eldest son was somehow in `trade`.

Luckily, so far our tastes have turned out not to be that rarified after all, and we have sold far more books than anyone ever thought that we would, so there is a moral in there somewhere…

Jon Downes,
Woolsery, North Devon
July 2010

CFZ PRESS

Other Books in Print

Weird Waters – The Mystery Animals of Scandinavia: Lake and Sea Monsters by Lars Thomas
Monstrum! By Tony `Doc` Shiels
CFZ Yearbook 2011 edited by Jonathan Downes
Karl Shuker's Alien Zoo by Shuker, Dr Karl P.N
Tetrapod Zoology Book One by Naish, Dr Darren
The Mystery Animals of Ireland by Gary Cunningham and Ronan Coghlan
Monsters of Texas by Gerhard, Ken
The Great Yokai Encyclopaedia by Freeman, Richard
NEW HORIZONS: Animals & Men *issues 16-20 Collected Editions Vol. 4* by Downes, Jonathan
A Daintree Diary -
Tales from Travels to the Daintree Rainforest in tropical north Queensland, Australia by Portman, Carl
Strangely Strange but Oddly Normal by Roberts, Andy
Centre for Fortean Zoology Yearbook 2010 by Downes, Jonathan
Predator Deathmatch by Molloy, Nick
Star Steeds and other Dreams by Shuker, Karl
CHINA: A Yellow Peril? by Muirhead, Richard
Mystery Animals of the British Isles: The Western Isles by Vaudrey, Glen
Giant Snakes - Unravelling the coils of mystery by Newton, Michael
Mystery Animals of the British Isles: Kent by Arnold, Neil
Centre for Fortean Zoology Yearbook 2009 by Downes, Jonathan
CFZ EXPEDITION REPORT: Russia 2008 by Richard Freeman *et al*, Shuker, Karl (fwd)
Dinosaurs and other Prehistoric Animals on Stamps - A Worldwide catalogue by Shuker, Karl P. N
Dr Shuker's Casebook by Shuker, Karl P.N
The Island of Paradise - chupacabra UFO crash retrievals,
and accelerated evolution on the island of Puerto Rico by Downes, Jonathan
The Mystery Animals of the British Isles: Northumberland and Tyneside by Hallowell, Michael J
Centre for Fortean Zoology Yearbook 1997 by Downes, Jonathan (Ed)
Centre for Fortean Zoology Yearbook 2002 by Downes, Jonathan (Ed)
Centre for Fortean Zoology Yearbook 2000/1 by Downes, Jonathan (Ed)

Centre for Fortean Zoology Yearbook 1998 by Downes, Jonathan (Ed)
Centre for Fortean Zoology Yearbook 2003 by Downes, Jonathan (Ed)
In the wake of Bernard Heuvelmans by Woodley, Michael A
CFZ EXPEDITION REPORT: Guyana 2007 by Richard Freeman *et al*, Shuker, Karl (fwd)
Centre for Fortean Zoology Yearbook 1999 by Downes, Jonathan (Ed)
Big Cats in Britain Yearbook 2008 by Fraser, Mark (Ed)
Centre for Fortean Zoology Yearbook 1996 by Downes, Jonathan (Ed)
*THE CALL OF THE WILD - Animals & Men issues 11-15
Collected Editions Vol. 3* by Downes, Jonathan (ed)
Ethna's Journal by Downes, C N
Centre for Fortean Zoology Yearbook 2008 by Downes, J (Ed)
DARK DORSET -Calendar Custome by Newland, Robert J
Extraordinary Animals Revisited by Shuker, Karl
MAN-MONKEY - In Search of the British Bigfoot by Redfern, Nick
Dark Dorset Tales of Mystery, Wonder and Terror by Newland, Robert J and Mark North
Big Cats Loose in Britain by Matthews, Marcus
MONSTER! - The A-Z of Zooform Phenomena by Arnold, Neil
The Centre for Fortean Zoology 2004 Yearbook by Downes, Jonathan (Ed)
The Centre for Fortean Zoology 2007 Yearbook by Downes, Jonathan (Ed)
CAT FLAPS! Northern Mystery Cats by Roberts, Andy
Big Cats in Britain Yearbook 2007 by Fraser, Mark (Ed)
BIG BIRD! - Modern sightings of Flying Monsters by Gerhard, Ken
*THE NUMBER OF THE BEAST - Animals & Men issues 6-10
Collected Editions Vol. 1* by Downes, Jonathan (Ed)
IN THE BEGINNING - Animals & Men issues 1-5 Collected Editions Vol. 1 by Downes, Jonathan
STRENGTH THROUGH KOI - They saved Hitler's Koi and other stories by Downes, Jonathan
The Smaller Mystery Carnivores of the Westcountry by Downes, Jonathan
CFZ EXPEDITION REPORT: Gambia 2006 by Richard Freeman *et al*, Shuker, Karl (fwd)
The Owlman and Others by Jonathan Downes
The Blackdown Mystery by Downes, Jonathan
Big Cats in Britain Yearbook 2006 by Fraser, Mark (Ed)
Fragrant Harbours - Distant Rivers by Downes, John T
Only Fools and Goatsuckers by Downes, Jonathan
Monster of the Mere by Jonathan Downes
Dragons:More than a Myth by Freeman, Richard Alan
Granfer's Bible Stories by Downes, John Tweddell
Monster Hunter by Downes, Jonathan

Fortean Words

The Centre for Fortean Zoology has for several years led the field in Fortean publishing. CFZ Press is the only publishing company specialising in books on monsters and mystery animals. CFZ Press has published more books on this subject than any other company in history and has attracted such well known authors as Andy Roberts, Nick Redfern, Michael Newton, Dr Karl Shuker, Neil Arnold, Dr Darren Naish, Jon Downes, Ken Gerhard and Richard Freeman.

Now CFZ Press are launching a new imprint. Fortean Words is a new line of books dealing with Fortean subjects other than cryptozoology, which is - after all - the subject the CFZ are best known for. Fortean Words is being launched with a spectacular multi-volume series called *Haunted Skies* which covers British UFO sightings between 1940 and 2010. Former policeman John Hanson and his long-suffering partner Dawn Holloway have compiled a peerless library of sighting reports, many that have not been made public before. Other books include a look at the Berwyn Mountains UFO case by renowned Fortean Andy Roberts and a series of forthcoming books by transatlantic researcher Nick Redfern.

CFZ Press are dedicated to maintaining the fine quality of their works with Fortean Words. New authors tackling new subjects will always be encouraged, and we hope that our books will continue to be as ground-breaking and popular as ever.

Haunted Skies Volume One 1940-1959 by John Hanson and Dawn Holloway
Haunted Skies Volume Two 1960-1965 by John Hanson and Dawn Holloway
Space Girl Dead on Spaghetti Junction - an anthology by Nick Redfern
I Fort the Lore - an anthology by Paul Screeton
UFO Down - the Berwyn Mountains UFO Crash by Andy Roberts

www.ingramcontent.com/pod-product-compliance
Lightning Source LLC
Chambersburg PA
CBHW071801090426
42737CB00012B/1906